D0305315

UK Price £4.95

AN ILLUSTRATED GUIDE TO

SPACE
WARFARE

Published by Salamander Books Limited
LONDON • NEW YORK

AN ILLUSTRATED GUIDE TO

SPACE

WARFARE

David Hobbs

A Salamander Book Credits

© Salamander Books Ltd.,
52 Bedford Row,
London WC1R 4LR,
United Kingdom.

ISBN 0 86101 204 6

Distributed in the United Kingdom by
Hodder & Stoughton Services,
P.O. Box 6,
Mill Road,
Dunton Green,
Sevenoaks,
Kent TN13 2XX.

Author: David Hobbs is the Director
of the North Atlantic Assembly's
Scientific and Technical Committee.
Trained initially as a physicist,
he has specialized in military
technology's influence on strategy,
and before taking up his current
position he was a researcher at
Aberdeen University's prestigious
Centre for Defence Studies.

Editor: Bernard Fitzsimons
Art Editor: Mark Holt
Designed by Stonecastle Graphics
Diagrams: TIGA

Photographs: The publishers wish
to thank all the companies and
other organizations who supplied
illustrations used in this book.

Typeset: The Old Mill, London
Printed in Belgium by Proost
International Book Production,
Turnhout

Contents

Introduction

SPACE HAS BEEN used for military purposes for more than 25 years. The United States launched its first successful reconnaissance satellite in 1960 and the Soviet Union followed suit in 1962. Since then, satellites have been developed to perform a wide variety of military tasks, ranging from electronic eavesdropping to communications, and space systems have played an ever-increasing role in terrestrial military operations.

The military exploitation of space is dominated by the Soviet Union and the United States. France, the United Kingdom and Italy have military communications satellite programmes, and China has launched some reconnaissance satellites, but these efforts are dwarfed by those of the superpowers. The Soviet Union launches about 100 satellites per year of which about 70 per cent are for exclusively military purposes and 15 per cent perform dual military and civilian functions.

The United States generally launches less than 20 military satellites per year. This disparity in launches of Soviet and American military satellites is misleading, however, since the latter have much longer lifetimes and are more versatile; the number of active satellites maintained by each nation is about 120. In the future, both nations will become even more active in space as the United States takes advantage of the Space Shuttle and as the Soviet Union brings into service a new heavy-lift rocket or reusable shuttle similar to the American one.

At present, the overwhelming majority of both superpowers' space effort is devoted to 'non-weapon' purposes such as intelligence gathering, communications and navigation. Space-based assets are 'force multipliers', allowing many traditional military missions to be conducted more efficiently, though as technology advances space systems are evolving into 'force enablers',

opening up new mission possibilities. To cite just a few examples, satellite systems now existing or in development will enable global control of forces, improved tactical communications, all-weather navigation, precise weapons delivery independent of range, and long-range naval target acquisition.

As reliance on military satellites has grown, so too has the awareness that the loss of satellites in wartime could be a crippling handicap. This has led to the development of techniques for disabling an opponent's satellites — by either destroying or jamming them — and has also led to the development of methods of protecting satellites from enemy action. So whereas space systems were formerly viewed as a means of aiding terrestrial military operations, space is now also seen as a theatre of war.

Moreover, war in space may not be confined to a battle between satellites. Strategic ballistic missiles travel through space on their way to their targets, and technological progress in many areas has now opened up the possibility of destroying these missiles in space. Both superpowers have large research programmes to assess the feasibility of a host of possible space weapons which might be suitable for this purpose. Before such weapons can be produced many enormous technical obstacles must be overcome, but progress is being made on many fronts and some spectacular results have already been achieved.

The decision to deploy space-based ballistic missile defences will depend on the outcome of research projects now in hand as well as on the political and strategic climate prevailing in the future. But the decision to use space for military purposes was made over thirty years ago with the result that space is already highly militarised. What follows is a description of the military use of space now, plus an account of where research is leading.

Space Warfare and Strategic Defence

ON MARCH 23 1983, US President Ronald Reagan made a speech in which he called upon the scientific community to 'give us the means of rendering . . . nuclear weapons impotent and obsolete'. Subsequently, the United States expanded and reorganized its research into ballistic missile defence and labelled the effort the Strategic Defence Initiative (SDI). This programme, popularly known as Star Wars, has given rise to heated controversy and has become one of the most important and contentious issues facing nations in both East and West.

There are many reasons why the SDI is such a controversial subject: fundamentally, some argue that the SDI could eventually provide a way of removing the threat of nuclear annihilation, while others argue that it could make a nuclear holocaust more likely. The Soviet Union takes the latter view, presenting the SDI as — among other things — an impediment to reductions in nuclear weapons and a threat to strategic stability. The Soviet Union, however, is itself conducting a research effort devoted to precisely the same goals. Indeed, both superpowers are investigating and developing a host of space warfare technologies ranging from satellite protection techniques to laser weapons.

Existing systems

Some space warfare systems already exist: the Soviet Union has an operational anti-satellite (Asat) system and the United States is developing one. These systems cannot strike satellites at high altitudes but they could be adapted to do so relatively easily, and research efforts may result in the production of new, more capable anti-satellite systems in the near future.

Both superpowers are concerned about existing Asat systems and about the potential appearance of even more potent forms of anti-satellite weaponry. Terrestrial military operations rely increasingly on satellites for navigation, communications, reconnaissance and a variety of other tasks, so their loss in a conflict could be a decisive disadvantage. It is very difficult, however, to see what can be done to reduce the anti-satellite threat. Some defensive measures can be and are being taken, such as hardening satellites to resist damage from Asat weapons, making them able to carry out evasive manoeuvres and making them difficult to detect, but these measures will not actually curtail the development of more capable anti-satellite weapons.

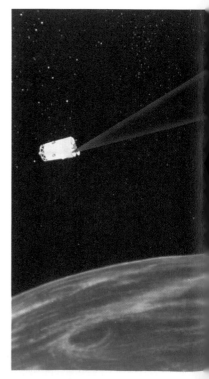

Above: The official badge of the Strategic Defense Initiative, widely known as Star Wars, symbolically depicts a shield rising into space.

Below: To many people the major proposal in the SDI is the development of laser weaponry in space, but the SDI in fact embraces many other concepts and technologies. The Soviet Union is also conducting a similar research programme.

An arms control agreement governing Asat weapons seems an attractive possibility, but many obstacles would have to be overcome. For a start, Asat systems are very easy to conceal, so it would be very difficult to guarantee compliance with an anti-satellite treaty. Consequently, the superpowers may choose simply to continue developing Asat weapons to ensure either that they can inflict equivalent damage on each other's military satellites, or that neither will use such weapons for fear of retaliation.

Although the issues surrounding Asat weapons may seem complicated, they are certainly less so that those surrounding ballistic missile defence, and to appreciate the latter it is necessary to examine the current philosophy of nuclear strategy.

Current strategy

At present, the superpowers guarantee their security with huge arsenals of nuclear weapons, including various kinds of ballistic mis-

siles, bombers and cruise missiles, but with the emphasis on ballistic missiles. These are essentially large rockets designed to hurl their lethal payloads into space at speeds of several kilometers per second: once in space, the relatively small warheads are dispensed towards their assigned targets. Once they are launched, unless they malfunction, ballistic missiles will assuredly reach their targets. The only exceptions are those which may fall foul of a defensive network around Moscow or some particularly high-performance Soviet surface-to-air missiles, but these exceptions are few and the United States has no equivalent defences.

The only way of substantially reducing the effects of a ballistic missile attack is to destroy the missiles before they are launched — a difficult task, because they are either buried in steel-reinforced concrete silos or are based under the sea aboard quiet, deep-diving submarines. It may be possible to destroy a significant proportion of land-based missiles as the accuracy of warheads steadily improves — though missiles could be launched before incoming warheads arrive — but the sea-based missiles are particularly secure, since it is extraordinarily difficult to locate and destroy a ballistic missile submarine. Consequently, the superpowers can rely for security on a strategy of mutual assured destruction (MAD) because, in effect, no matter how determined an attack, no matter what an adversary does, there is nothing that can prevent a massive

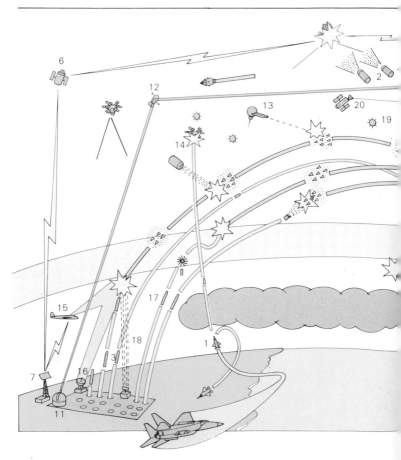

retaliatory strike. On that basis, according to current strategic thinking, an adversary will not launch a nuclear attack for fear of an equally devastating response.

Enhancing survivability

As the accuracy of ballistic missiles increases and as anti-submarine warfare techniques improve, there is a need to improve the survivability of ballistic missiles to ensure a viable retaliatory capability. Means of enhancing survivability include increasing both the capabilities of missile-carrying submarines and the range of their missiles, so that the submarines can hide in larger areas of ocean while remaining in range of their targets, and operate closer to friendly forces for protection. The survivability of land-based missiles can be improved by further strengthening of their silos, but only up to a point: once an adversary's missiles can achieve almost direct hits, no amount of hardening will suffice. Various other methods of protection have been proposed such as making ballistic missiles mobile or constructing active defences around silos to intercept incoming warheads.

The ABM Treaty

The last course of action, however, is limited by the Anti-Ballistic Missile (ABM) Treaty, which permits the United States and the Soviet Union to deploy only 100 missile interceptors each — not a great deal of protection for the thousand or more ballistic missiles each side has, particularly when an attack could consist of around 10,000 warheads.

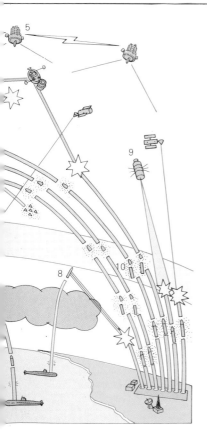

Left: Theoretical scenario, with each side trying to blind the other by hitting early-warning and communications satellites using air-launched (1) and orbiting Asats (2). Land-based ICBM launchers (3) are accompanied by some SLBMs (4). Early-warning satellites (5) track all launches and data is relayed to ground control (7) via communications satellites (6). Engagements start as soon as possible, since post-boost phase hits kill missile and warheads in one blow; submarine-launched nuclear-pumped X-ray lasers (8) and orbiting laser battle stations (9) start destroying attacking missiles. Surviving missiles dispense warheads and decoys (10); warheads are engaged by ground-based lasers (11) using orbiting mirrors (12), electro-magnetic rail-guns (13) and satellites armed with small rocket interceptors (14); surviving warheads are tracked by airborne (15) and ground-based (16) sensors. In the final stages ground-based interceptors (17) are launched to destroy warheads, ending with terminal 'swarmjet' defences (18). Both sides also use mines (19) and other Asat weaponry (20).

Right: Survivability of nuclear forces can be improved by denying hostile powers exact knowledge of ICBM locations. In the 1970s the USAF tested the deployment of MX missiles in C-5 aircraft; the missile was extracted by parachute as shown here. Trials were successful but the system was not adopted.

Below: A variety of secure basing modes were proposed for the MX missile system in the 1970s. One involved a network of tunnels through which moved missile transporters capable of breaking through the overhead concrete and earth cover to enable the missile to be launched.

The notion underlying all these improvements to ballistic missile survivability is the preservation of the MAD strategy, and the ABM Treaty itself is interpreted as an acknowledgement of the principle that both superpowers guarantee their security through mutual vulnerability. This is why ballistic missile defence research is so controversial: it implies that the MAD principle could be overthrown, effectively revolutionizing nuclear deterrence strategy.

Before exploring the nature of this revolution in strategic doctrine it must be pointed out that there is considerable disagreement about whether ballistic missile defences on a large scale will prove technically feasible and economically affordable. For the moment, it is convenient to set this debate aside in order to examine the theoretical implications.

Although the precise form of large-scale ballistic missile defences cannot yet be established, it is expected that several layers would be needed, each thinning out an attack at a different stage. The first layer

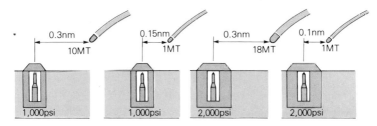

Above: The counter-force role requires ICBMs to destroy the enemy's ICBM force which, in all current cases, is located in hardened silos. Kill probability depends upon warhead yield and accuracy, the latter being expressed in terms of circular area probable (cep), the radius in nautical miles (nm) of a circle in which 50 per cent of the warheads will land. As shown here, the cep is the more important factor: to achieve a 90 per cent kill probability on a 1,000psi silo requires a 10MT weapon at 0.3nm, but if accuracy can be doubled (ie, the cep reduced to 0.15nm), only a 1MT weapon is required. Similarly, a 2,000psi silo will be destroyed by an 18MT warhed at 0.3nm, while a 1MT warhead needs an accuracy of better than 0.1nm.

might intercept missiles shortly after launch, and successive layers would engage the surviving warheads as they approached their targets. It is by no means certain whether such defences would have beneficial or adverse strategic consequences, but the general nature of the argument is clear.

The case for missile defences is that they would remove or at least substantially reduce the threat of nuclear annihilation, permitting a movement toward a strategy of mutual assured survival. Strategic stability would be guaranteed because there would be little to gain from a nuclear exchange, not because there would be everything to lose. Also, security would depend not on an adversary's perceptions of the damage he would suffer in a retaliatory strike, but on the capability of defences.

Such defences could not be emplaced overnight, of course; rather, there would be a gradual increase in defensive capabilities as successive systems were deployed. During this transition phase, and even if de-

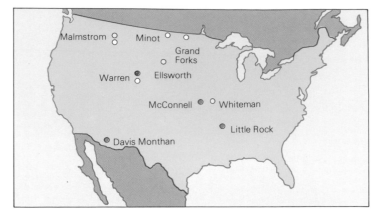

Above: ICBM sites in Continental USA (CONUS), showing their central location, although withdrawal of the final Titan IIs in 1987 will reduce the overall dispersion. The 50 Peacekeepers authorized by the US Congress will be located in selected Minuteman silos. (Red=MX bases, blue Titan II, yellow Minuteman II and white Minuteman III).

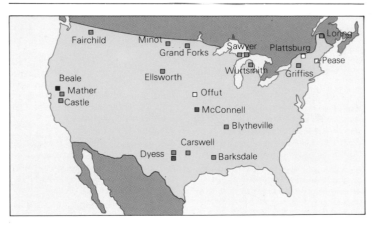

Above: Strategic aircraft bases in CONUS, including those allocated to the B-1B. These aircraft, together with the essential airborne tanker fleet of KC-135s and KC-10s, constitute the air-breathing leg of the US strategic triad. (Bases shown in red will house the new B-1Bs, blue the remaining B-52 force, yellow the EC/RC-135, black the SR-71 and white the FB-111.)

fences were never perfected, sufficient attacking missiles could be destroyed to preserve a retaliatory capability. An adversary might be able to estimate what proportion of his missiles would be intercepted, but not which he could not execute a successful first strike and would not attempt one.

Also, an adversary placing greater emphasis on bombers and cruise missiles in an attempt to circumvent defences would be helping stability because these systems are slower than ballistic missiles and can not preempt the launch of a counterattack. An additional benefit of ballistic missile defences would be

Above: Soviet strategic bomber and ICBM sites, plus interceptor bases, showing the two focuses, one in the west and the second in the east, with further sites along the Trans-Siberian Railway.

Squares indicate aircraft bases (yellow for bombers, black for interceptors); circles are ICBM sites (red SS-11, blue SS-13, yellow SS-17, black SS-18 and white SS-20).

that accidental launches could be intercepted, as could small strikes from nations which may acquire nuclear strike capabilities in the future.

The case against ballistic missile defences is that no matter how effective they could be, they could always be circumvented by bombers and cruise missiles so that the threat of nuclear annihilation would still be present. Furthermore, promoting a sense of security could lead the superpowers to exercise less self-restraint in the conduct of their affairs and could precipitate a war which would ultimately be devastating since the threat from bombers and cruise missiles would remain.

Orbital weakness

Another argument — one which is both technical and strategic — is that the space-based elements of one side's defensive systems would be vulnerable to attack from the other's space weapons. The orbiting components would follow predictable paths and would be larger and fewer than the warheads in a ballistic missile attack, so in times of crisis there would be a standing temptation to try to eliminate the other side's systems before the adversary

tried to do the same thing. There would be a considerable incentive to strike first, leaving the adversary stripped of his defences while facing an opponent who remained well defended. Also, the assembly of a defensive network by one side could exert pressure on the other to launch an attack before losing the chance.

Another strategic consideration is the effect that missile defences would have on the relationship between the United States and its allies. There is concern that if the United States and the Soviet Union were well protected, they might be less concerned about a conflict developing in Europe. Even though many defensive technologies being studied could be applied in Europe, concern persists because if defences neutralized the missile threat, other force imbalances would become even more significant. The Warsaw Pact currently has considerable superiority in both nuclear bombers and conventional forces for use in Europe, and missile defences on both sides would make these imbalances more significant and more worrying.

Whether such defences would affect the US commitment to European defence is a political not a

Below: Some of the concepts being explored in the US BMD programme include ground- and space-based elements. Some may be realised in the near-term, but others, such as directed

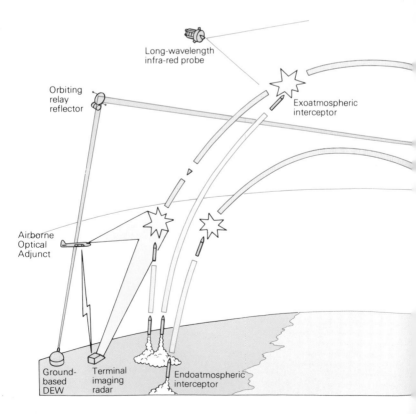

Long-wavelength infra-red probe

Orbiting relay reflector

Exoatmospheric interceptor

Airborne Optical Adjunct

Ground-based DEW

Terminal imaging radar

Endoatmospheric interceptor

Left: A picture that will bring memories to older artillerymen of the great railway guns of years gone by. In its search for more survivable deployment options the Soviet Union is considering railways for the fifth-generation SS-X-24 ICBM. The missile will be silo-deployed initially (1986), with rail deployment probably following in 1988/89, and will carry 10 MIRVs.

technical judgement. A well-defended United States might be even more ready to commit itself to Western Europe because it would have less to fear from Soviet nuclear retaliation. Conversely, the United States might decide to leave Western Europe to its own devices, though it should be pointed out that the United States has given repeated assurances to the contrary.

Another aspect of the debate is the effect of missile defences on the missile forces of the United Kingdom and France, which are large enough to penetrate moderate defences but much smaller than those of the superpowers, and which might be rendered ineffective by US and Soviet missile defences. Much depends on the time-scale for defensive deployments. By one or two decades into the next century, the two West European deterrent forces could be replaced by other systems such as cruise missiles and bombers in the course of a normal modernization schedule. If deployments took place earlier, however, expensive replacements would have to be found before existing and planned forces reached the end of their budgeted lives.

energy weapons and orbiting reflectors, need long-term advances in technology.

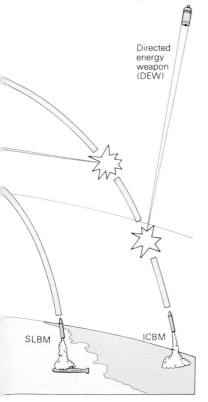

Directed energy weapon (DEW)

SLBM

ICBM

Canadian concerns

Canada too has some specific concerns. Technical considerations may well suggest that some elements of a US defensive system would be most efficiently located on Canadian soil, possibly providing protection for the host country but also possibly increasing the likelihood of Canadian territory being attacked. In addition, there could be domestic opposition to providing facilities for US defensive systems.

However unappealing some of these consequences may appear for the United States and its allies, it should be remembered that not only the United States is investigating ballistic missile defence technologies. The Soviet Union also has a vigorous research effort in progress and still less appealing is the prospect of the Soviet Union deploying defences without the United States being able to follow suit, a situation whose strategic implications are obvious.

Deployment consequences

Analysis of the potential consequences of deploying ballistic missile defences is clearly hypothetical, since no decision has been made to proceed with such deployments, nor can it be until the outcome of current research is known. In the United States a decision may be reached in the early 1990s, while there is no in-dication of when the Soviet Union might be technically able or politically prepared to proceed with large-scale deployments, and the strategic consequences would depend on both the technical capabilities of any systems deployed and the strategic environment prevailing at the time.

For instance, the superpowers might agree to deploy certain types of defences in order to ensure the survivability of their land-based missiles as part of an agreed arms control formula which would enhance strategic stability rather than undermine it. Eventually, they might even agree to move toward a strategy of mutual assured survival supported by improved defences and reduced ballistic missile, bomber, and cruise missile forces. Admittedly the elimination of nuclear offensive forces — and more besides — would be a preferable formula but since the

---

Below: The US ballistic missile warning system. Satellites detect Soviet ICBM launches within 90 seconds and instantly warn ground stations on Guam and in Australia. BMEWS then tracks and identifies missiles, while PARCS identifies RVs and predicts impact sites. SLBM warnings come from satellites and Pave Paws, while the FPS-85 covers the threat from the south.

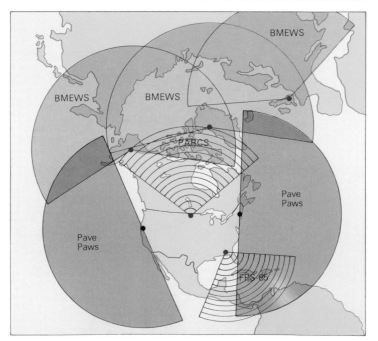

superpowers are unlikely to trust each other, and since other nations may develop nuclear forces, they may prefer the reassurance of a mixture of defensive and offensive forces.

Destabilizing potential

On the other hand, defensive deployments could take place as part of an arms race, with each side seeking to thwart the other's defences by building up its ballistic missile, bomber, and cruise missile forces. This would be destabilizing and extremely expensive. It is theoretically possible that effective defences could still be constructed to defeat an attack even in the face of a vast multiplication of the threat, but it would be technically and financially even more demanding than the already formidable task of countering today's ballistic missiles.

These arguments show that the value of constructing ballistic missile defences in the future will depend upon many related factors — not only technical appraisals of effectiveness and cost, but just as importantly, the nature of East-West relations when deployment is considered, the existence of an agreed arms control framework for deployment, and the attitudes of allies. As to whether such defences will prove technically feasible, economically affordable and strategically desirable, it is impossible to judge, despite assertions to the contrary. If it were indeed obvious that defences were utterly impractical, then the United States and the Soviet Union would presumably not be spending so much money on defensive research, nor would the Soviet Union be so vigorously opposed to American expenditure on such research.

Below: The USSR has the world's most comprehensive system for early warning of ballistic missile and air attacks. The ballistic missile EW system includes a launch-detection satellite network, over-the-horizon radar and a series of large phased-array 'Hen House' radars located along the periphery of the USSR. The US considers the Krasnoyarsk radar to breach the ABM Treaty.

'Hen House'
New radars
'Dog House'/'Cat House'
Krasnoyarsk radar

Th Space Rac

ON OCTOBER 3, 1942, German scientists launched an A-4 rocket which travelled a distance of 118 miles (190km) and reached an altitude of over 50 miles (80km), causing one of the scientists involved in the A-4's development to remark that 'today the spaceship was born'. The A-4, however, was not to be used to explore space: instead, it was armed with a ton of high explosive and used primarily to attack London and Antwerp. After World War II the United States and the Soviet Union rounded up the missile scientists and equipment and employed them in their own efforts to build missiles and space launchers.

Much to America's surprise, it was the Soviet Union which first succeeded in launching an artificial satellite: on October 4, 1957, Sputnik 1 was launched into orbit from the Baikonur cosmodrome, and for three weeks the satellite transmitted signals to Earth. Then, on November 3, Sputnik 2 was launched with a dog on board.

Until that point American space and missile programmes had been proceeding at a relatively leisurely pace, with a variety of research efforts and feasibility studies being conducted often in competition rather than cooperation with each other, but these Soviet successes shook American confidence and led to a dramatic expansion and streamlining of space and missile programmes. The United States was justifiably concerned about the potential military advantage implied by Soviet space achievements: if the Soviet Union could place an object into space, it could equally use missiles to attack the United States, and there was a widespread feeling that the US could no longer defend itself.

Explorer 1

At the end of January 1958, American self-confidence was partly restored by the successful launch of Explorer 1, a scientific satellite, but the damage had been done: the United States considered itself to be behind the Soviet Union in space and missile technology and felt an urgent need to catch up.

The American effort to catch up was expensive and fraught with embarrassing failures. During the two years following the launch of Sputnik 1, the Soviet Union conducted five space launches, all successful: the United States attempted 30 laun-

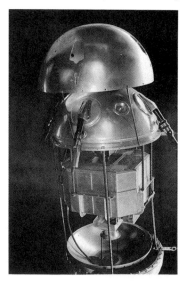

Left: The start of it all. Sputnik 1 was launched by the USSR on October 4, 1957, to the amazement of the world and the consternation of the USA, which had no idea the Soviets were technically so advanced.

Above right: Explorer 1 on the launch pad. Orbited on January 31, 1958, it worked perfectly, doing much to restore American morale and proving instrumental in discovering the van Allen belt.

Right: Some four months after the USA had launched Explorer 1, weighing 31lb (14kg), the USSR launched Sputnik 3, seen here, which weighed 2,068lb (938kg). The Soviet satellite remained in orbit until April 6, 1960.

ches, of which 19 failed. American Discoverer photographic reconnais-sance satellites proved particularly troublesome. The first launch attempt took place on February 28, 1958, but the satellite failed to achieve a usable orbit, and 11 subsequent attempts also failed in one way or another; not until August 10, 1959, was Discoverer 13 launched on the first successful photographic reconnaissance mission. Despite the numerous setbacks, by the early 1960s the United States had achieved some notable successes, including the first reconnaissance, meteorological, navigation, early-warning, and communications satellites. The Soviet Union, meanwhile, achieved more spectacular firsts by launching the first lunar probes and, on April 12, 1961, by placing the first manned space vehicle into orbit, when Vostok 1 with cosmo-

Below: In the sad chapter of early US setbacks the Discoverer series managed to do especially badly. Discoverer 1 (seen here), launched on February 28, 1958, failed to reach a satisfactory orbit, and the following 11 were also unsuccessful for various reasons. Finally, with Discoverer 13, came success.

naut Yuri Gagarin on board completed one orbit and returned safely to Earth. The first American in space, Alan Shepard, Jr, was launched into space — though not into orbit — in a Mercury capsule on May 5, 1961, and Virgil 'Gus' Grissom experienced a similar flight in July 1961.

Soviet follow-up

Then, on August 6, 1961, the Soviet Union launched Vostok 2 with Gherman Titov on board. Vostok 2 orbited the Earth 17 times in a mission lasting over 25 hours. Not until February 20, 1962, did the first American orbit the Earth with John Glenn aboard the Mercury capsule *Friendship 7*, which orbited the Earth three times in just under five hours.

The US advantage in military satellites was short-lived, and during the course of the 1960s the Soviet Union launched satellites to perform similar missions. Of more concern was the Soviet lead in manned spaceflight and the heavy missile capability it implied: space launch vehicles were often modified missiles which clearly could be used to deliver nuclear warheads anywhere in the world, and there was no defence against them. Fears were also expressed about bombardment from space by orbiting nuclear weapons, and there was the further possibility that anti-satellite weapons would be developed to deprive the United States of its orbital eyes and ears. In October 1960, for instance, while campaigning for the Presidency, John F Kennedy stated that 'we are in a strategic space race with the Russians and we have been losing . . . Control of space will be decided in the next decade. If the Soviets control space they can control Earth, as in the past centuries the nations that controlled the seas dominated the continents'. Indeed, many of the concepts and arguments which appear in current debates about space warfare first appeared during the late 1950s and early 1960s. Ideas of space denial, space control, and space as 'the new high ground' appeared, and research began on anti-satellite and ballistic missile defence weaponry.

The United States was also concerned about its international prestige and national self-esteem. Soviet successes in manned spaceflight called American technological superiority into question just as the launch of Sputnik 1 had done. Consequently, in 1961, President Ken-

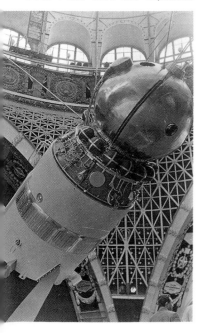

Above: Soviet cosmonaut Major Yuri Gagarin completed one orbit and returned safely to Earth on April 12, 1961. It was a notable first by a very brave man.

Left: One of the very successful Soviet Vostok manned spacecraft. Close inspection reveals the essential simplicity of the design, with a spherical manned capsule for strength.

nedy declared that the United States would launch a manned mission to the moon before the end of the decade, a project intended to re-establish American self-confidence and technological status.

The 1960s were a decade of intense space activity with world attention focussed primarily on Soviet and American manned missions. After John Glenn's orbital flight, three more Mercury missions took place, the last being in May 1963. Soviet Vostok missions ceased the following month with the flight of Vostok 6, which carried Valentina Tereshkova, the first woman in space. Vostok 6 passed within three miles (5km) of Vostok 5 which had been launched two days earlier and stayed aloft until shortly after Vostok 6 returned to Earth. Manned space missions then ceased for over a year to be resumed in October 1964 with the flight of Voskhod 1, based on the Vostok design but carrying three cosmonauts. Voskhod 2 was launched in March 1965 carrying two cosmonauts one of whom, Alexei Leonov, passed through Voskhod 2's airlock to become the first man to 'walk' in space. Then Soviet manned space flights stopped again for over two years, during which time the United States conducted 10 missions using new two-man Gemini space capsules, designed to spend up to two weeks in orbit and much more sophisticated than their Mercury predecessors. The first manned Gemini mission took place in March 1965, and the last in November 1966. Several Gemini capsules docked with unmanned Agena vehicles, demonstrating a technique that would be essential for future missions to the moon.

Right: As the US programme gathered pace in the mid-1960s the Gemini spacecraft — a two-man capsule with much greater capabilities than the earlier Mercury — entered service. Here Ed White becomes the first US astronaut to walk in space during the Gemini 4 mission in June 1965. Alexei Leonov had carried out the first ever space-walk three months earlier.

Left: The great American triumph as two astronauts walk on the surface of the Moon, July 21, 1969. Taken by the mission commander, Neil Armstrong, this picture shows Ed Aldrin beside a package of scientific experiments, with the lunar lander in the background.

Below: Ed Aldrin extracting a package from its stowage bay: another picture by Neil Armstrong. With this mission the goal set by President Kennedy in 1961 of placing a man on the Moon by the end of the decade was achieved; it has yet to be matched by the USSR.

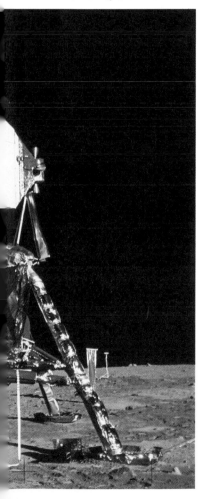

The following year began with a tragedy for spaceflight, when three astronauts — Virgil Grisson, Edward White and Roger Chaffee — were killed in an Apollo capsule which caught fire as they were rehearsing launch procedures. Then, in April, the first Soviet Soyuz mission ended in disaster as the capsule became entangled in its parachute and crashed, killing cosmonaut Vladimir Komarov. No further manned missions were attempted until October 1968, when Apollo 7 was launched on an 11-day mission to test hardware and techniques to be used on missions to the moon. The Soviet Union also launched Soyuz 3 with one cosmonaut on board, and this capsule docked with the unmanned Soyuz 2 capsule. Finally, in December 1968, Apollo 8 carried three astronauts on a six-day mission around the moon.

Space rescue rehearsal

In January 1969 Soyuz 4 and Soyuz 5 were launched within one day of each other. Soyuz 4 carried only one crew member, while Soyuz 5 carried three, and these vehicles performed the first docking of two manned spacecraft, following which two crew members from Soyuz 5 transferred to Soyuz 4 for reentry, rehearsing in effect a space rescue.

In March 1969 Apollo 9 was launched to test the lunar module that would be used to place astronauts on the moon. Two astronauts boarded the lunar module and conducted manoeuvres using its descent engines, then jettisoned the descent stage and fired the ascent stage to return to the Apollo command module. Two months later, Apollo 10 carried out essentially similar procedures, but this time around the moon, a rehearsal for a moon landing that took the lunar module astronauts to within nine miles (15km) of the moon's surface.

Then, in July 1969 Apollo 11 carried out the first manned moon landing, astronauts Neil Armstrong and Edwin 'Buzz' Aldrin descending to the surface in their lunar module while Michael Collins remained in the command module in orbit. On July 21, 1969, Neil Armstrong left the lunar module to become the first man

to walk on the moon, thus achieving the American goal.

In October 1969, the Soviet Union launched Soyuz 6, 7 and 8 on successive days, and the three capsules conducted extensive manoeuvres, closely approaching each other but not actually docking. This was the first time that three manned space vehicles were in orbit simultaneously. Finally, the last manned space mission of the 1960s, in November 1969, saw Apollo 12 achieve another moon landing, again collecting samples of moon rock and leaving scientific instruments behind.

Although manned space activity attracted most attention during the 1960s, unmanned scientific and military programmes also flourished, and a host of scientific spacecraft were launched by both superpowers.

Left: One of the first military communications satellites in orbit, the US Army's Courier was launched in October 4, 1960.

Below: The Skybolt air-launched ballistic missile, end result of the Bold Orion programme which included the first Asat test.

The majority were placed into orbit around the Earth to study the planet and to perform astronomical observations, while probes were launched to the moon, Venus and Mars.

Dramatic growth

Military space activity grew dramatically during the 1960s, so that by the end of the decade both superpowers were using military satellites for photographic and electronic reconnaissance, navigation, communications, early-warning and meteorology. The annual total of military satellite launches grew from about 15 in 1960 to over 100 in 1967, and though numbers declined slightly toward the end of the decade, an average annual rate of over 100 military launches has been maintained ever since. The distribution of launches changed substantially during the 1960s, however, with the average number of US military satellite launches declining from a peak of 55 in 1966 to 29 in 1969, largely as a result of the increasing endurance of reconnaissance satellites. In contrast Soviet military satellite launches steadily increased, from only five in 1962 to 66 in 1969.

Both superpowers also developed anti-satellite weapons in the 1960s. The first anti-satellite test actually occurred in 1959 when the United States launched a two-stage missile from a B-47 aircraft during tests for a project known as Bold Orion which was investigating the feasibility of an air-launched ballistic missile. The last test in a series of 12 was used to demonstrate a satellite interception technique, using a scientific satellite, Explorer VI, as the target. The missile passed within four miles (7km) of Explorer VI but despite this encouraging result, the United States favoured a ground-launched anti-satellite system.

Subsequently, two such weapons were developed. The first used a Nike-Zeus missile equipped with a nuclear warhead, but was abandoned in favour of a system incorporating a nuclear-armed Thor mis-

Above: A production Spartan anti-ballistic missile is launched from the Kwajalein Missile range. Spartan was the exoatmospheric element of the Safeguard system and was designed to intercept missiles at ranges of up to some 465 miles (748km) at speeds of Mach 10. It was armed with a nuclear warhead.

sile, which had a superior altitude capability. This Thor system, also known as Program 437, became operational in 1964 and remained nominally operational until 1975.

US ABM developments

The United States also continued to develop anti-ballistic missile weaponry during the 1960s, although no system was actually deployed. Work had begun on the Nike-Zeus missile as a short-range missile interceptor in the late 1950s, but although this missile was deemed suitable for anti-satellite purposes it was not judged capable of missile interception, so a new, faster version labelled Nike-X and subsequently named Sprint was developed, along with a longer-range missile interceptor named Spartan. The operational concept called for Spartan to engage missiles outside the atmosphere and Sprint inside it, and since accuracy was constrained by the technology of the period both were to have nuclear warheads.

Above: Cancelled in 1963, the Boeing Dynasoar was to have been boosted into space by Titan III, and, after its mission, reenter the atmosphere and glide to a landing. Its similarity to the more recent Soviet orbiter (page 155) is striking.

In 1959 research had begun on a more ambitious missile defence system, the Ballistic Missile Boost Interceptor (Project BAMBI), intended to be a system of hundreds of satellites, each equipped with small heat-seeking missiles to be launched down onto Soviet ballistic missiles as they rose from their launch pads. This project was deemed too technically demanding and vulnerable to countermeasures, so research ceased in 1963. There was other research on a variety of laser anti-satellite proposals, but the projects did not involve a great deal of effort or resources and did not result in the development of any weapon systems.

Also cancelled in 1963 was the Dynasoar project. This was envisaged as a small, reusable manned vehicle which would perform strategic reconnaissance, satellite inspection and intercontinental bombardment. Dynasoar — short for dynamic soaring — would be launched into space by a conventional rocket and would then bounce along the upper atmosphere before recovering to a normal runway, but rising costs and alternative ways of performing the missions led to its cancellation.

Next, work began on another manned military space system, the Manned Orbiting Laboratory (MOL), which was to consist of a Gemini capsule attached to a cylindrical laboratory section which would

Left: USAF Manned Orbiting Laboratory (MOL) with a modified Gemini capsule attached. Advances in unmanned spacecraft technology led to cancellation in 1969.

Below: These impressive missile containers seen during a Red Square parade in 1965 were a mystery for some time. They eventually proved to be ABM-1B 'Galosh' anti-ballistic missiles of the type used around Moscow.

allow astronauts to work in a shirt-sleeve environment and conduct military surveillance and reconnaissance. Other activities were to include communications, navigation and meteorology, but the MOL was overtaken by technology as advanced unmanned satellites were developed to perform many of these tasks. In addition, the defence budget was being stretched by the Vietnam War so in 1969 the MOL was cancelled.

Soviet activity

As noted above, Soviet military space activity also increased dramatically during the 1960s, and as well as launching a growing number of military satellites the Soviet Union also began work on an anti-satellite weapon. The first clear test of this system took place in 1968, but some of the required techniques such as in-orbit rendezvous had already been demonstrated. The Soviet anti-satellite system was not nuclear-armed; instead, it was equipped with a conventional explosive which sent a blast of metal fragments towards its target, and after four tests in the late 1960s, two of which were judged successful by the United States, the system was believed to be operational.

The Soviet Union also developed a nuclear-armed ballistic missile interceptor known as the 'Galosh', which was deployed around Moscow along with the necessary radar systems.

The first major space event of the 1970s was nearly a disaster. In April 1970, as Apollo 13 approached the moon for what was to have been the third US manned landing, an oxygen tank exploded, and only through some remarkable improvisation was it possible to use the lunar module's supplies and engines to return Apollo 13 safely to Earth. Between January 1971 and December 1972 a further four Apollo lunar landings were successfully achieved, whereupon the Apollo programme of lunar exploration ceased.

American manned space activity continued with the launch of Skylab in May 1973. Much bigger than any previous manned spacecraft, Skylab provided more than 12,000 cubic feet

(350m³) of workspace and weighed over 90 tons. Almost immediately after Skylab was placed into orbit it became evident that something was wrong, and it transpired that the thermal shield had been ripped off, carrying with it one of the solar panels and causing another to jam only half open.

Skylab repairs

Launch of the first Skylab crew was immediately postponed. Engineers on Earth appraised the damage and considered ways of repairing it, while special tools were constructed and the astronauts carried out under-water rehearsals of the necessary repairs. Eventually, on May 25, two weeks late, the crew was sent up in an Apollo capsule, and — with great difficulty — successfully docked with Skylab, eventually rendering it operational. A parasol sunshade was erected to replace the lost thermal shield and the jammed solar panel was freed, and the repairs were so successful that the crew were able to carry out 80 per cent of the work originally scheduled for the mission. The first crew remained aboard for 28 days and was succeeded by two further crews, the second being launched on July 28, 1973 and staying aboard for 59 days while the third and final crew was launched on 16th November 1973 for a stay of 84 days.

The Skylab missions provided a wealth of scientific data. Materials processing experiments were performed, an astronaut manoeuvring unit was tested, and an enormous amount of data was collected about astronomical phenomena, solar physics and Earth resources. In addition, valuable observations were made of Soviet missile centres and activities along the Sino-Soviet border.

Joint operation

The final American mission of the 1970s, in July 1975, was part of a joint American-Soviet operation. An Apollo capsule and a Soyuz capsule docked with each other and the crews worked together on a variety of experiments for almost two days.

Soviet manned space activity continued throughout the 1970s with 26

Above: Jack Lousma, Skylab 3 pilot, taking part in a successful two-man spacewalk to deploy a sunshade to protect the damaged space station from the glare of the Sun. This took place during the second visit to the space station.

Right: The Apollo flights seemed to have become routine when Apollo 13 developed a serious fault following an explosion some 205,000 miles (329,845km) from Earth. The astronauts improvised various devices to help them return safely; this was the carbon dioxide purger.

successful Soyuz launches and one aborted mission. Most of the Soyuz missions were associated with the Salyut space stations, the first of which was launched on April 19, 1971. The Salyut space stations were smaller than Skylab but were equipped to perform a great deal of scientific and military work.

Soyuz 10 conducted the first docking with Salyut 1 but the crew did not attempt to enter the space station and the vehicles separated after five and a half hours. Soyuz 11, launched on June 6, 1971, docked with Salyut 1 and the crew transferred to the space station, spending 23 days aboard before returning to Earth. A faulty valve caused the Soyuz vessel to decompress during reentry, however, and the crew died. Modifications to the Soyuz design were necessary, so no further crews were able to occupy Salyut 1 before it burned up in the Earth's atmosphere on October 11, 1971. Salyut 2, launched on April 3 1973, broke up in orbit, and it is believed that two further Salyut launches attempted shortly afterwards were unsuccessful.

Salyut 3 was launched on June 25, 1974, on a primarily military mission.

The crew of Soyuz 14 occupied the space station for 15 days and left the station to function automatically. Soyuz 15 failed to dock with Salyut 3, but the space station continued to take reconnaissance photographs which were jettisoned and returned to Earth before it finally reentered after seven months in orbit.

Salyut 4, 5 and 6, launched in December 1974, June 1976 and September 1977 respectively, demonstrated a variety of new Soviet capabilities, including automatic docking of Soyuz capsules and Progress supply vessels and automatic refuelling. Some Soyuz crews also stayed in space for record-breaking lengths of time, five long-term missions on Salyut 6, for instance, lasting 96, 140, 175, 185 and 75 days. Salyut 6 also had docking ports at

Below: In 1975 the two space powers undertook the joint Apollo-Soyuz Test Project (ASTP). This involved much preliminary cooperative work and culminated in a docking operation in space. Here the US astronauts toast their Soviet colleagues in borsch.

each end, greatly adding to the space station's flexibility. Like Salyut 3, Salyut 5 was dedicated to mainly military missions, notably photographic reconnaissance.

Manned space missions during the 1970s made great contributions to science, but other activity was no less important, the decade seeing the launch of the first probes to the outer planets with the American Pioneer and Voyager missions and the first soft landings on Mars and Venus. Satellites were also used to study the Earth and the sun, and to collect various other forms of astronomical information, while commercial communications satellites, a novelty in the 1960s, became commonplace in the 1970s.

Military satellites continued to follow the trends established in the 1960s, American launches declining from about 30 in 1970 to an average of less than 20 for the rest of the decade as their capabilities gradually improved. Soviet launches, on the other hand, increased from more than 60 in 1970 to an average of almost 90 for the rest of the decade. There were two main reasons for the larger number of Soviet launches — an increasing use of communications satellites, and the practice of increasing reconnaissance capabilities by using satellites in greater numbers rather than by substantially increasing their endurance and ability to return data to Earth.

1970s Asat tests

As for anti-satellite weapons, the United States conducted three tests in 1970, all using variants of the Thor missile, but tests then ceased and the anti-satellite system, Program 437, was decommissioned in 1975.

The United States was no longer convinced of the utility of its anti-satellite systems. One of the main reasons for deploying them was the threat of bombardment by orbital nuclear weapons, and in 1967 the Outer Space Treaty had been signed, banning the deployment of such systems. The Soviet Union, however, continued its anti-satellite programme and during the 1970s conducted 13 anti-satellite tests. Some of these used a radar homing device

and were unsuccessful, but most of those involving an infra-red homing device succeeded.

The 1970s also saw the beginning and the end of American anti-ballistic missile deployments. The 1972 Anti-Ballistic Missile Treaty placed a strict limit on ABM deployments, and an agreement in 1974 restricted deployments still further. The United States did not feel that the proposed Safeguard system in North Dakota, with its Sprint and Spartan interceptors, afforded worthwhile protection, and indeed did not consider that defensive technology was capable of

Right: The Sun's flaring corona prepared from data supplied by NASA's Solar Maximum mission (1980). Such scientific projects have greatly expanded our understanding of the Cosmos.

Below: The surface of Mars as seen by the US Viking lander. In 1986 the USSR launched a manned space station whose tasks may include acting as a base for a manned Mars flight.

E

HAO SMM CORONAGRAPH/POLARIMETER
DOY 103   UT= 1416 POL=0

offering realistic protection from ballistic missile attack, so the Safeguard system was dismantled. The Soviet Union, however, retained its defensive 'Galosh' system around Moscow primarily, it was believed, to counter the smaller potential nuclear attacks from nations such as the United Kingdom, France and China. The ABM Treaty did permit research into defensive systems, however, and both superpowers continued to conduct such research.

The 1980s saw Salyut 6 still in orbit. This space station was regularly refuelled, re-provisioned and partly refurbished by visiting crews and unmanned Progress supply vessels, so successfully that it remained in orbit until the end of July 1982. Before then, though, in April 1982, Salyut 7 was launched, and this space station, unlike its forerunners, was designed for both civil and military functions.

Some problems were experienced with Salyut 7, but repairs were successfully carried out and at the end of 1985 the station was still in orbit and still active. The 1980s also saw the reintroduction of three-man Soviet crews using a modified Soyuz capsule known as the Soyuz T. (Since the Soyuz 11 disaster, only two-man crews had been used.)

The space shuttle

The most significant development in manned spaceflight, however, was undoubtedly the introduction of the American space shuttle, whose first launch, on April 12, 1981, was the first manned American space mission for almost six years. By the end of 1985, the full fleet of four shuttles was operational, providing unprecedented access to space.

At the beginning of 1986, however, the shuttle *Challenger* exploded shortly after launch, killing the seven crew aboard in the worst space disaster in history. Despite this tragedy, and earlier comparatively minor setbacks, the shuttle remains an enormously flexible space vehicle, serving as a launcher, an experimental platform and a satellite repair or retrieval system. As such, it will play a key part in the US space plans until the end of the century.

Above: The US space shuttles were so reliable that their flights began to seem routine until the January 1986 disaster. This

launch — NASA's eighth and Challenger's third — on August 30, 1983, was the first to begin in darkness (local time 02:32): ominously, it was later found that the mission had come close to disaster through erosion of a solid rocket booster nozzle.

Military Satellites
Early Warning and Attack Assessment

INTERCONTINENTAL ballistic missiles (ICBMs) take only about 30 minutes to reach their targets and submarine-launched ballistic missiles (SLBMs) might take only 10 minutes, depending on the launch and target locations. These short flight times make it essential to detect and assess a nuclear attack as rapidly as possible. Accordingly, the superpowers employ early-warning satellites which use infra-red sensors to detect the tremendous heat from a ballistic missile's exhaust, seconds after launch. These satellites also monitor missile tests and satellite launches.

The United States maintains three early-warning satellites in geostationary orbit. Launches from the Soviet Union and China are monitored by a satellite over the Indian Ocean, and satellites over the Pacific Ocean and South America monitor SLBM launches.

Each of these Defense Support Program (DSP) satellites carries a Schmidt telescope 11ft 11in (3.63m) long with a 3 ft (0.91m) aperture. At the telescope's focus is an array of 2,000 lead sulphide infra-red detectors, each of which scans an area 3¾ miles (6 km) across. Orientation is maintained by spinning the satellite at about 6rpm around the Earth-pointing axis, and the telescope is offset from this axis by 7.5°, producing a conical scanning pattern. By plotting an infra-red source over several scans it is possible to determine whether the source is stationary — a forest fire, for instance — or moving. A missile can be identified within a minute of initial detection, which usually occurs as it breaks through cloud cover.

In 1975 an early-warning satellite was temporarily blinded and there

Right: Satellites play a critical role in monitoring ICBM launches, principally by using infra-red sensors to detect the intense heat of the rocket efflux. This is a US Defense Support Program early warning satellite ready for launch.

Below: Impression of a DSP satellite in orbit. At the focus of the 11ft 11ln (3.63m) long telescope is an array of 2,000 lead sulphide (PbS) infra-red detectors, each scanning an area 3.75 miles (6km) across.

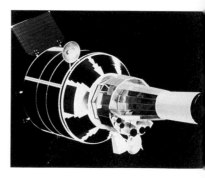

was concern that a Soviet ground-based laser was responsible, though the US Department of Defense (DoD) later stated that the blinding was caused by an intense fire at a broken gas pipeline. However, the DoD does believe that the Soviet Union has two ground-based lasers that could have anti-satellite capabilities.

The HALO programme
Fears about laser blinding, and the prospect of more powerful laser weapons, are taken so seriously that the United States has embarked on

several programmes to increase the survivability of early-warning satellites, and plans are also in hand to improve their sensing capability. The High Altitude Low Observable (HALO) programme is intended to produce laser-resistant high-resolution sensors which stare constantly at their target areas instead of scanning them periodically. Ablative materials are also to be used to make early-warning satellites less vulnerable to laser weapons. These features should be integrated into the next-generation Satellite Early-Warning System (SEWS), which is

to be deployed in the early 1990s.

Before the next generation of early-warning satellites enter service, however, a number of near-term measures are being taken to improve the survivability of the present generation. Replacement DSP satellites will be hardened against nuclear radiation and will be equipped with devices to close off the optics to prevent damage from lasers. In addition, more powerful satellite-to-ground data transmitters and satellite crosslinks will be used. At present data is transmitted to ground stations at Alice Springs, Australia, and

Buckley Field, Colorado, and is then relayed to command and control centres through the Defense Satellite Communications System (DSCS). The ground stations are vulnerable to attack but new measures will permit the use of small, mobile ground receivers, thus enhancing the survivability of the overall early-warning system.

Nuclear detection

Early-warning satellites are also equipped with devices to detect nuclear explosions. In peacetime these sensors help verify compliance with the treaty banning atmospheric nuclear tests, while in a nuclear war they would enable rapid damage assessment to be carried out. By 1988, however, this function will be performed by the Integrated Operational Nudet (nuclear detonation) Detection System (IONDS). IONDS sensors will be fitted to Navstar navigation satellites and will be able to provide precise data on the location, altitude and yield of nuclear explosions.

Virtually all Soviet satellites are known simply by a number in the Cosmos series, and early-warning satellites are no exception. Cosmos 520, launched in September 1972, was the first to be identified and gradually, over the next ten years, a nine-satellite early-warning system was assembled. The satellites are

Below: A USAF Atlas-Agena launcher blasts off from Cape Canaveral on July 20, 1965, with the third pair of TRW-built Vela satellites on board.

Right: This picture of a Vela satellite in final assembly was taken in the 1960s and shows the components and wiring of a bygone electronic age.

spaced at regular intervals around a highly elliptical, semisynchronous orbit inclined at 62°. Geostationary orbits are difficult to achieve from the Soviet Union's northern launch sites, but the orbit used by Soviet early-warning satellites provides good coverage of the Northern hemisphere and each satellite can view American ICBM sites for up to six hours per 12-hour orbit.

In 1983, three replacement early-warning satellites were launched to take the place of those which had reached the end of their operational lives. In 1984, however, seven new early warning satellites were launched, replacing virtually the entire network.

Below: A Vela satellite, one of 12 which monitored deep space for Test Ban Treaty violations from 1963 until replaced by the IMEWS system.

Surveillance and Reconnaissance

Surveillance and reconnaissance satellites — often known as spy satellites — are used to obtain information about a multitude of military activities. Surveillance is a relatively regular monitoring activity, whereas reconnaissance is a search for specific intelligence, possibly of a more urgent nature. Although the tasks are different, they are increasingly combined on a single satellite platform equipped with several sensors.

Photographic surveillance and reconnaissance satellites use optical, infra-red and, possibly, radar techniques to obtain highly detailed pictures of areas of interest. The information produced can be of incalculable value, enabling the strengths and locations of forces to be determined with considerable precision and weapons under construction or testing to be observed in remarkable detail. The resolution of satellite photographs is so fine that, for instance, the size of an aircraft engine's air intake can be measured, so a weapon's performance can be estimated while still in the prototype stage, perhaps years before it is fielded. Military exercises and actual conflicts can also be monitored: both superpowers observed the Falklands and Iran-Iraq wars using satellites, and such satellites are also the primary means of verifying compliance with treaties.

Big Bird

Probably the best-known spy satellite in the American inventory is Big Bird. Big Bird can perform both wide-area surveillance and 'close-look' high-resolution reconnaissance with multi-spectral scanners. Its cameras can identify objects as small as 12in (30cm) across and its film is processed on board, scanned by an optical system and transmitted to receiving stations on Earth. These transmitted images are of poorer quality than photographs, so if exceptionally fine detail is required, Big Bird will jettison film capsules to be recovered in mid-air by specially equipped HC-130 Hercules aircraft based in Hawaii.

Big Bird satellites are placed into low-altitude sun-synchronous orbits so that they pass over their targets at the same time each day. The orbit can take the satellite as low as 100 miles (160km), where atmospheric drag would normally cause reentry in about a week. Big Bird, however, has rocket motors which are used periodically to nudge the satellite back into position, extending its life to about 200 days.

The United States also uses two other photographic reconnaissance satellites designated KH-8 and KH-9 (for Key Hole). Their existence only became public knowledge in 1983 and details are sparse, but they are known to be film-return types. Their orbits are low and lifetimes are

Above: Big Bird reconnaissance satellite launch by a Titan IIID. Big Bird carries out both wide-area surveillance and high resolution detailed reconnaissance missions.

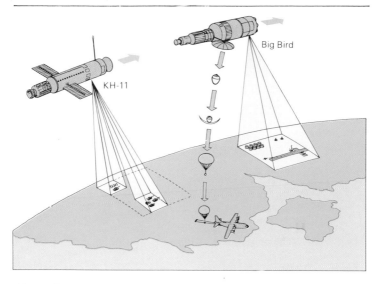

Big Bird

KH-11

Above: Big Bird's high resolution cameras give broad views from which targets are selected for discriminatory analysis. It is backed up by the top secret

KH-11 (Key Hole) digitized reconnaissance satellite, which transmits high quality TV images back to Earth for near real-time detailed analysis.

Right: Big Bird's near real-time system develops films on board and transmits them to Earth; results are rapid but of less than photographic quality. There is a back-up system which returns film capsules to Earth for mid-air recovery by C-130s with special equipment.

limited — KH-9 can probably remain aloft for about four months — so their use is confined to photographing intelligence targets of the highest priority.

Production of Big Bird, KH-8 and KH-9 has ceased, and increasing use is being made of KH-11 which, unlike its predecessors, does not return film. Instead, it uses a digital imaging device reportedly able to achieve resolutions comparable to Big Bird's despite KH-11's higher operational altitude. The higher altitude extends orbital life to perhaps more than two years, and the imaging system makes information available almost instantaneously. KH-11 has multispectral and infra-red sensors and, according to some reports, a radar sensor capable of penetrating cloud cover.

KH-12 is reportedly under development with improvements over KH-11 likely to include higher resolution imaging and, almost certainly, radar sensors..

Soviet reconsats

The first Soviet photo-reconnaissance satellite mission took place in 1962, and for most of the 1960s two types of photo-reconnaissance satellite were used. These first and second generation models had mission lives of about eight days and were replaced by the more capable third and fourth generation models which are still used.

Above: A film capsule ejected by a satellite returns through the atmosphere by parachute to be captured in mid-air by a specially modified C-130, a system used for more than ten years.

The third generation was introduced in 1968 with the launch of Cosmos 208. Like its predecessors, this model returns to Earth at the end of its mission complete with its cameras and film. This type is generally used for low resolution surveillance missions lasting approximately 14 days. Recently, a new operational capability was observed, however. Cosmos 1587 was launched on August 6, 1984, and was given a one-day checkout before being shut down for 10 days. At the end of this dormant period the satellite was reactivated and then performed a standard, 14-day reconnaissance mission. Cosmos 1613, launched on November 29, 1984, also followed this pattern.

The fourth generation photo-reconnaissance satellite, first introduced in 1975 with the launch of Cosmos 758, weighs about 15,200lb (6,900kg), almost 4,400lb (2,000kg) more than the third generation types, and is based on the Soyuz manned space capsule. This model is highly manoeuvrable and is used for high-resolution close-look missions. A film return system is used consisting of six recoverable film capsules

which are jettisoned and returned to Earth periodically. When first introduced, this satellite had a mission life about 30 days, but improvements such as the replacement of chemical batteries with solar panels have extended missions duration to about 60 days.

In 1984 the Soviet Union introduced a completely new form of photo-reconnaissance satellite, using digital imaging techniques and relaying data through geosynchronous communications satellites. Cosmos 1543, Cosmos 1552 and Cosmos 1608 were believed to be of this type and they operated for 26, 173 and 33 days respectively. This new model will add considerably to the Soviet Union's intelligence-

Above: The view from the Hercules ramp as the film capsules's parachute is snared by the mid-air pick-up. The HC-130H aircraft will now head rapidly for its base so that film processing and intelligence assessment can begin as soon as possible.

gathering capability by transmitting near real-time imagery for long periods, and as a result of its introduction Soviet reconnaissance satellites logged 982 missions operation days in 1984, exceeding the previous year's record by 152 days, despite the fact that in both 1983 and 1984, the same number — 27 — of photo-reconnaissance satellites were launched.

The Soviet Union also uses its Salyut space stations for photo-reconnaissance. Although all missions perform a mixture of civilian and military functions, Salyuts 3 and 5 were of a predominantly military character. The 32ft 10in (10m) focal length camera, ostensibly for solar observations, was better suited for reconnaissance, and the lower orbit of Salyuts 3 and 5 also implied this function. Furthermore, the crew were military officers, code words were used to conceal the meaning of conversations between the crew and ground-control, and telemetry was transmitted in forms previously associated with reconnaissance satellites.

In April 1982 Salyut 7 was launched on a primarily civilian mission, but the station exhibited a capability which will prove extremely useful in sustaining military tasks. In March 1983, Cosmos 1443 docked automatically with Salyut 7, which at that time was unmanned. The Cosmos vehicle carried a large propellant load which was transferred to Salyut. This ability to refuel Salyut automatically, thereby increasing its orbital life and manoeuvrability, will be particularly useful for extending low-orbit reconnaissance missions, which require frequent nudges to stay aloft.

Elint satellites

Photographic reconnaissance is by no means the only method of intelligence gathering from space. Electronic intelligence (elint) satellites — also known as ferrets — have many applications, such as locating radio transmitters, eavesdropping on communications and monitoring the telemetry from missile tests. A radar's purpose can be established by examining pulse width, pulse repetition frequency, transmitter frequency and modulation. This type of information is useful for studying peacetime military activities and it greatly facilitates the planning of wartime operations. Known radar sites can be attacked — or avoided

— by strike aircraft and effective electronic countermeasures can be developed to jam radio communications and deceive radars.

At least two types of American elint satellites are known to be in service. One is a subsatellite ferret usually launched along with Big Bird but subsequently boosted into a circular orbit at an altitude of about 300 miles (500km), or in some cases about 900 miles (1,500km). The principal purpose of subsatellite ferrets is to conduct general surveys, detecting new radar sites and monitoring changes at known ones.

The other type of ferret is known as Rhyolite and two of these are maintained in geostationary orbits, one above the Horn of Africa and the other over the Indian Ocean. Rhyolites bristle with antennas, the largest being over 65ft (20m) in diameter, and on-board electronics filter out background noise and jamming signals. It is believed that one of Rhyolite's primary missions is to monitor missile launches from the Tyuratam and Plesetsk launch sites. The Rhyolites now in operation were launched in the early 1970s and may already have been supplemented by their eventual successor, known as Aquacade. Virtually no information about Aquacade has been published but in view of the remarkable progress made in electronics during the last decade there can be no doubt that Aquacade will be more sensitive and powerful than Rhyolite.

Soviet systems

The Soviet Union employs various sorts of elint satellite but, as is generally the case, little is known about them. There is a series of eight ferrets operating in a 400-mile (650km) orbit with an inclination of 81.2° whose task is to identify and pinpoint military radars and radio transmitters, while other ferrets are launched into a variety of orbits.

In September 1984, a completely new Soviet elint satellite, Cosmos 1603, was launched. Although few details of this spacecraft are avail-

Left: The Soviet Salyut 7 space station at Baikonur Cosmodrome prior to its April 1982 launch. Salyut 7 was refuelled in space by Cosmos 1443 in a totally automatic operation.

Below: This satellite picture of a 'Blackjack' and two Tu-144s at Ramenskoye is of poor quality. Those available to intelligence experts are specially processed to give greater clarity.

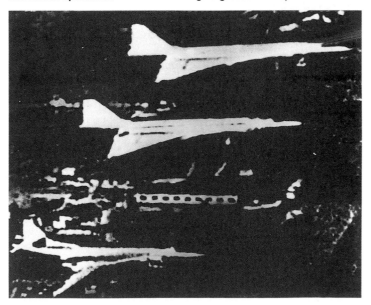

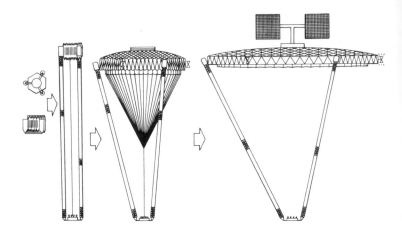

Above: The largest of Rhyolite's numerous antennas is 65ft (20m) in diameter. One technique for deploying such a big structure would be to orbit the antenna in packaged form (left). Booms would then extend to deploy the backing for the reflector (centre), with lanyards holding the reflector membrane clear of the expanding truss structure. Electro-static charges would then draw the membrane rearward, stretching its surface into a paraboloid, while the booms remained in place to hold the antenna feed at the focus of the newly-made dish.

Below: The reason for the massive antenna dish on the Rhyolite satellite is that it has the high gain needed to detect the tiny amounts of energy in the very weak side-lobes of the narrow beams of microwave communications links operating deep within the USSR. Such faint signals could not be detected by earlier satellites which had only low-gain antennas.

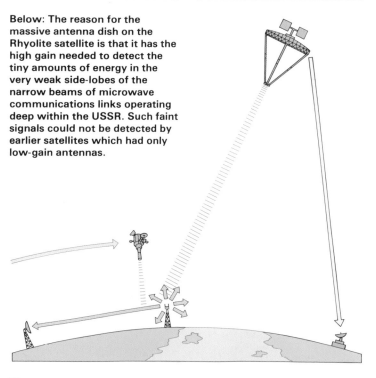

Above: US satellite picture of the USSR's top secret Plesetsk cosmodrome, located south of Archangel at 62° 43′ N, 40° 18′ E, and first identified publicly by a British schoolmaster.

able, it appears to be the largest military satellite ever launched by the Soviet Union and was initially placed into a low orbit at an altitude of about 110 miles (180km), inclined at 51.6°. After a day or so the satellite's propulsion system was used to shift the orbit to an altitude of 530 miles (850km) inclined at 66.6°, and shortly afterward the inclination was changed to 71° while altitude remained the same. These manoeuvres made tracking very difficult and the United States lost the satellite several times. Cosmos 1603's final orbit could have been achieved much more efficiently and this has given rise to speculation that the Soviet Union was testing its ability to avoid anti-satellite weapons.

Ocean surveillance

Satellites fitted with elint equipment and various other sensors are also used for ocean surveillance. The main American satellite system for this purpose, White Cloud, consists of three clusters of satellites spaced at 120° intervals around a 680-mile (1,100km) orbit inclined at 65.3°. Each cluster comprises three small satellites, each weighing only a few tens of kilograms, dispersed from one 'parent' satellite. The clusters orbit in a formation which allows the data from infra-red and millimetre-wave sensors on each satellite to be collated, providing information on surface vessels more than 1,800 miles (3,000km) away.

Work is taking place with a view to fitting similar satellites with radar, colour scanners, scatterometers and sensitive infra-red detectors. These will provide detailed information about surface vessels and may even enable the detection of warm water trails from nuclear powered submarines. Equipment for this is being developed as part of the National Oceanic Satellite System, which will also provide ocean data for civilian applications, while another programme, known as the Integrated Tactical Surveillance System (ITSS), will provide the techniques for monitoring naval vessels, aircraft and missiles.

Soviet ORSATS

The Soviet Union operates elint ocean reconnaissance satellites (EORSATs) at altitudes of about 280 miles (480km) with inclinations depending upon mission. They frequently work in pairs and have low-thrust engines to maintain correct height and spacing. EORSATs are used to monitor communications and radar emissions from naval missions.

Radar-equipped ocean-surveillance satellites (RORSATs) are used to detect and track surface vessels and these, too, frequently operate in pairs. RORSATs generally orbit at altitudes of around 155 miles (250km), and even with thrusters to maintain altitude their orbital lifetimes are limited to about 70 days. However, these satellites are powered by nuclear reactors containing around 110lb (50kg) of slightly enriched uranium so at the end of a RORSAT's life the reactor section separates and is boosted into a 560-mile (900km) orbit, while the rest of the satellite falls to Earth.

On at least three occasions this process has failed and the reactor has reentered. In one case, on January 24, 1978, reactor fragments from Cosmos 954 fell over the Northwest Territories of Canada, necessitating a six-million-dollar clean-up operation. After that incident ROR-SAT missions were suspended until 1981. In February 1983, however, after another malfunction, the reactor from another RORSAT, Cosmos 1402, reentered the Earth's atmosphere over the South Atlantic. Missions were again suspended until 1984.

EORSATs and RORSATs are known to be used to track American carrier task forces in particular, but they are also used to monitor targets of opportunity such as the Falklands War.

Right; The US Teal Ruby system is part of the High Altitude Low Observable (HALO) programme, developing space-based infra-red mosaic arrays for detecting and tracking air activity over hostile or foreign airspace. It will be shuttle-launched.

Right: In the hunt for SSBNs it is first necessary to establish ambient criteria by monitoring weather (1), sea state (2), oceanographic data (3) and thermal variations (4), and identifying unwanted sources (eg, fish) (5) and relating solar activity to Earth's magnetic field variations. Elimination of merchant ships (7) is by voluntary reporting (8) and satellites (9). SSBN tracking starts with satellite photography (10) as they leave port and electronic monitoring (11, 12). ASW aircraft (13) use various detectors: MAD, sonobouys (14), thermal devices (15) and forward-looking infra-red (16); Rapidly Deployed Surveillance System (17) is also air-delivered. Surface ships use sonar (18) and towed arrays (19), as do attack submarines (20). Passive devices include Sound Surveillance Systems (SOSUS) (5) and seabed coils which monitor electrical field variations

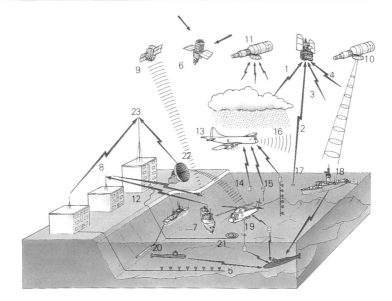

(21). Finally, traces of a submarine's passage can be detected by Over-the-Horizon (Backscatter) radar (22). All this information is fed back to a huge computer (23) which analyzes and correlates the data from the various sources.

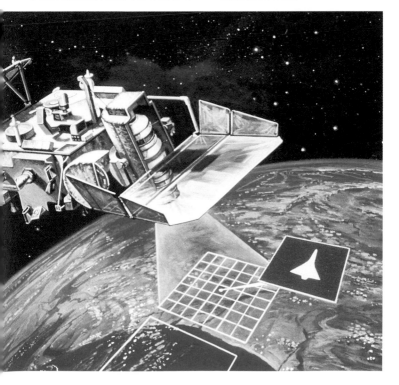

Communications

The role of communications satellites is obvious, though it is less well known just how heavily military forces rely on them: over 70 per cent of all American overseas military communications are relayed by satellite, for instance. And their importance is growing. As technology advances, sensors and weapons operate at greater ranges and the complexity of warfare increases dramatically, so that real-time long-range communications are more essential than ever.

The main US military satellite communications systems are the Defense Satellite Communications System (DSCS), the Air Force and Fleet Satellite Communication Systems (AFSATCOM and FLTSATCOM) and the Satellite Data System (SDS).

DSCS is the DoD's primary network for long-haul high-volume communications between major facilities. One network, DSCS-II, consists of four geostationary satellites plus two in-orbit spares which can handle 1,300 voice channels or 100 megabits per second. The DSCS-III satellites being phased in will provide a greater data-handling capacity along with high ECM resistance, hardness against EMP and propulsion systems to manoeuvre in case of attack.

USN and USAF comsats

FLTSATCOM, whose primary purpose is to provide naval communications, also consists of four geostationary satellites. More than 900 relay links are in use on ground stations, surface vessels, submarines and naval aircraft. The system became operational in 1980 and provides almost global coverage, but is still not adequate for the US Navy's needs, and three additional FLTSATCOM satellites have been scheduled for launch before 1989.

AFSATCOM is not a satellite system in its own right, but comprises communications packages carried on other satellites such as FLTSATCOM, SDS, DSCS and others. AFSATCOM terminals are used by E-4B airborne command

posts, RC-135 reconnaissance aircraft, Strategic Air Command bombers, TACAMO aircraft and various ground stations such as ICBM command posts.

SDS is a three-satellite network in highly elliptical orbits to fill the polar 'gaps' which geostationary systems leave in communications. This system is also used to relay data from KH-11 reconnaissance satellites.

These military satellites do not fully satisfy the US armed forces' communications requirements and the shortfall is particularly acute for the US Navy, so the DoD leases channels on commercial communications satellites. Six channels leased on three Comsat General MARISAT satellites form a system designated Gapfiller, used mainly for two-way communications with ships. This system is reaching the end of its operational life and the US Navy is now using channels on the Hughes Aircraft Corporation's Leasat system.

By 1990 the Military Strategic Tactical and Relay (Milstar) system

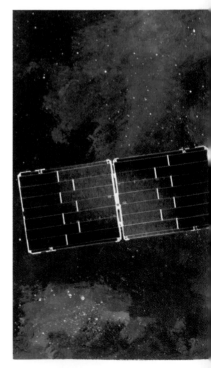

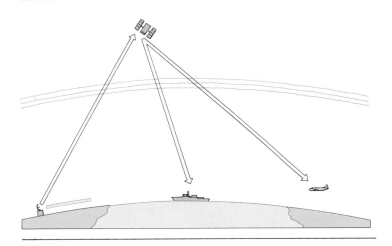

Above: VHF and UHF links (yellow) are essentially line-of-sight, and use either terrestrial or space-based relays to extend their range. HF, however, uses ground-wave (blue) for short ranges or reflections (red) off the ionosphere for transmissions over longer ranges.

Below: The latest US military communications system — DSCS-III — uses very sophisticated satellites which are EMP hardened and able to manoeuvre in space to evade Asat attacks. They also incorporate highly classified ECCM techniques.

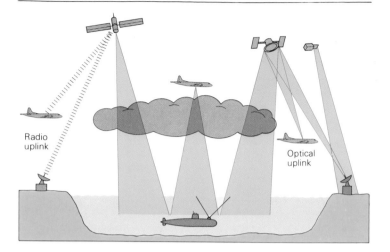

Above: A major problem with SSBNs is that of reliable and survivable communications. One method now being tested is the use of lasers, particularly in the visible blue-green part of the electromagnetic spectrum. One way of achieving this is to use a space-based laser to relay radio signals. If the satellite can not carry sufficient power then a space-based mirror might be used to reflect the signal from a ground-based laser, or an aircraft could be used as the vehicle for the laser station.

should be operational, complementing and eventually superseding FLTSATCOM, AFSATCOM, DSCS and, possibly, SDS. Milstar will be the first system to employ the millimetre-wave region, which will provide an enormous communications capacity. Four satellites plus one in-orbit spare will be in geostationary orbits and three will be in highly elliptical polar orbits. An unspecified number of spares will also be placed in supersynchronous orbits at altitudes of 110,000 miles (177,000km). Eventually, all Milstars may be placed in supersynchronous orbits to reduce their vulnerability. The satellites will be hardened against EMP and lasers, and will also be cross-linked by laser and other systems.

Submarine communications
Despite the sophistication of all these satellites, none can solve a specific communications problem which has plagued strategic planners for more than 20 years: communicating with nuclear submarines. At present, sub-

marines have to approach the surface to receive command instructions, making them more liable to detection. Work is under way on extremely low frequency (ELF) transmissions which can penetrate the ocean depths, but these would require huge antennas and have very poor data transfer rates. An alternative approach is to communicate using a satellite-borne blue-green laser. Blue-green laser light penetrates the atmosphere and sea water to a considerable depth. Tests using an airborne laser to communicate with a submarine have been very encouraging, so laser communication satellites may soon remove the chink in submarines' armour.

The Soviet Union currently employs three main types of military communications satellite. The Molniya I operates in an elliptical orbit inclined at 62.8° and a series of these are stationed at 45° intervals. The orbital period is 12 hours and each satellite remains visible from the Soviet Union for about eight hours per orbit. Molniya I satellites are used for direct point-to-point communications.

The second type, used for tactical communications, are launched eight at a time into 930-mile (1,500km) circular orbits inclined at 74°. Twenty-four are needed to ensure global coverage and 30 or more are generally operational, ensuring redundancy and resistance to jamming. Used for real-time command and control communications over the Soviet Union and Eastern Europe, these satellites also have a store-dump facility whereby a transmission is recorded, then played back and re-transmitted for global communications.

The third type of satellite is a store-dump type which operates at 500 miles (800km). Only one is active at a time and it is believed that this type is used to collect data from remote sensors and spies.

Not yet operational, but believed to be in development, is a system known as GALS, a four-satellite geo-stationary system along the lines of the American DSCS.

Below: A US Navy Fleet Satellite Communications (FLTSATCOM) satellite being lowered from an environmental test chamber. There are four such satellites in geo-stationary orbit, with another three scheduled for launch by 1989 to form a comprehensive global coverage system.

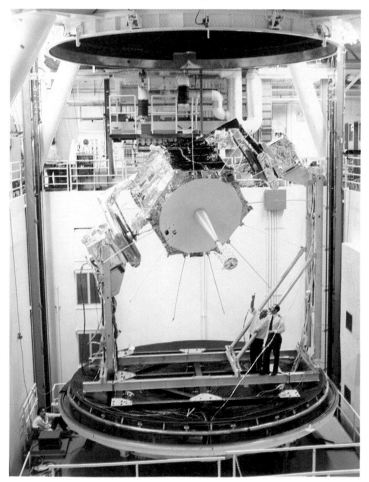

Below: Communications satellites (comsats) provide military forces with a very cost-effective means of passing traffic, whether it is voice, telegraph, facsimile or data. Such systems link national command authorities to army commanders in the field, to surface ships and submarines at sea, and to aircraft. The terminal equipment involved has been reduced in size from the original large stations with huge dishes which required fleets of trucks to move them to today's which can be totally accommodated in a single half-ton trailer.

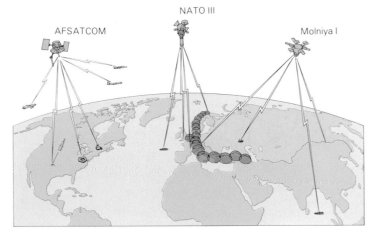

AFSATCOM

NATO III

Molniya I

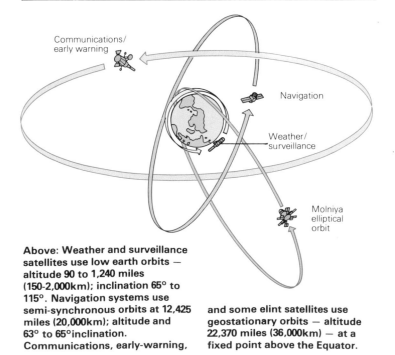

Communications/early warning

Navigation

Weather/surveillance

Molniya elliptical orbit

Above: Weather and surveillance satellites use low earth orbits — altitude 90 to 1,240 miles (150-2,000km); inclination 65° to 115°. Navigation systems use semi-synchronous orbits at 12,425 miles (20,000km); altitude and 63° to 65°inclination. Communications, early-warning, and some elint satellites use geostationary orbits — altitude 22,370 miles (36,000km) — at a fixed point above the Equator.

The two superpowers do not have a monopoly on military communications satellites. The United Kingdom, France and NATO also operate such systems, and the UK was the first nation to operate a geostationary military communications satellite, Skynet 1, purchased from Ford Aerospace in the US and launched in November 1969. Skynet 1B and Skynet 2A failed to achieve their desired orbits but Skynet 2B was successfully placed into a geostationary orbit over the Seychelles in November 1974.

Future European comsats

Skynet 3 was cancelled and Britain relied on American, NATO and leased commercial channels to supplement Skynet 2B. Two Skynet 4 satellites are now planned for launch, however, and there is a possibility of a third. Skynet 4 satellites will support strategic, maritime and land tactical operations and will be resistant to nuclear radiation and electronic jamming. The system is optimized for the British armed forces' requirements, providing relatively low data-rate communications for a widely dispersed fleet, and a higher volume capacity for mobile land forces in Europe.

France does not as yet operate dedicated military communications satellites but uses special channels on its commercial Telecom satellites. Telecom 1A and 1B occupy geostationary orbits at 8° W and 5° W re-

spectively, providing comunications over an area from the West Indies to the Reunion Islands; the military segment is designated Syracuse and provides communications facilities for naval and land forces. An expansion of the Syracuse system is planned but it had not been decided by the end of 1985 whether Syracuse II will consist of a dedicated satellite system or be combined with civilian space assets.

NATO currently employs three satellites designated NATO III A, B and C, plus NATO III D, launched in November 1984 to act as an in-orbit spare, to provide secure and permanent communications links between national command authorities and the Alliance's command structure. The NATO III satellites are fully interoperable with DSCS III and Skynet 4 satellites. By 1985 NATO was investigating how to satisfy its future communications requirements and was expected to procure four new satellites to be placed in orbit in the early 1990s. An off-the-shelf system was favoured to minimize costs, and the leading contenders were DSCS III and Skynet 4.

Below: The NATO IIIB satellite undergoing testing in an anechoic chamber; it was launched in January 1977. The system is invaluable to NATO in peace but its survivability in war could be compromised by future Soviet Asat developments.

Navigation

The first American navigation satellite, known as Transit, was originally developed to enable Polaris submarines to update their navigation systems accurately in all weathers. A series of Transit satellites in 680-mile (1,100km) circular orbits transmit oscillating radio signals which can be picked up by small receivers. Position can be calculated to within 500ft (150m) by measuring the Doppler shift in the transit signals and combining this with knowledge of the satellites' orbital motion. Since becoming operational in 1964 the system has been regularly improved. Most recently, Nova satellites, incorporating more powerful transmitters and with a better ability to compensate for orbital disturbances, have been used in the Transit network. Time lapses between position updates vary between 30 minutes at high altitudes and 100 minutes near the equator. The system is used by several thousand civilian and military operators world-wide.

Some years ago, military planners realized that technology had progressed to the extent that a navigation satellite system could be produced which would provide extremely precise position and velocity information to users equipped with a very small receiver. Consequently, the Navstar Global Positioning System (GPS) programme was initiated.

Navstar satellites

GPS will consist of 18 Navstar satellites placed at regular intervals around three orbital rings, each inclined at 63° to the Equator. The satellites' altitude will be about 12,500 miles (20,000km) with a period of 12 hours, and each satellite will transmit in two different codes, one for military and the other for civilian use. The former will enable a user to establish position in three dimensions to within 50ft (15m) and velocity to within a few centimetres per second. The military signal will be encrypted and highly resistant to

Below: Navstar Global Positioning System satellites continuously transmit identity, precise position in space and current time (according to on-board clocks). Receivers note message arrival time, work out how long the signal took to arrive and hence their own distance from the satellite. Measurements from three spacecraft give a positional fix; the fourth indicates errors in the user's reference clock. No ship, aircraft or tank with this system need ever be geographically lost.

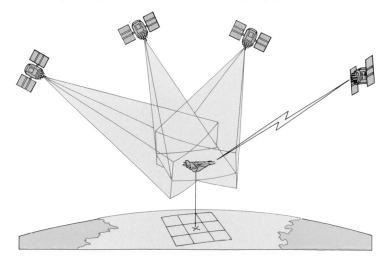

jamming, and user terminals will be man-portable and will remain passive. Navstar prototype satellites have been tested with considerable success and the full system should be operational by the end of 1988.

The applications for Navstar are almost limitless, and it is anticipated that the system will service 20,000 users. Ships, aircraft, missiles and ground forces will be able to establish position with unprecedented accuracy using passive receivers, and weapons will be able to achieve remarkable precision over any range. SLBMs such as Trident will be able to achieve the same accuracy as ICBMs; cruise missiles will be able to navigate without emitting any signals; air support for ground forces will be improved; minesweeping will be more efficient, since swept areas will be mapped more precisely; reconnaissance satellites will be able to provide more accurate information about potential targets; and aircraft will be able to rendezvous for in-flight refuelling without radar or radio assistance.

Soviet navigation satellites follow the American pattern almost exactly. One system, comparable to Transit, is in 620-mile (1,000km) polar orbits and is maintained by about four launches per year. The Soviet Union's latest system, Glonass, is almost a carbon copy of Navstar and will have both military and civilian users. The satellites will be in three orbital planes at an inclination of about 63° and an altitude of 12,500 miles (20,000km). The 12 satellites planned will not provide complete coverage, leading to speculation that Glonass military receivers might be designed to process Navstar's signals (possibly the 'coarse' civilian frequency).

Below: A Navstar GPS Satellite undergoes tests in the space simulation chamber before being launched into orbit. There will be 18 satellites at regular intervals in three orbital rings, and the information they provide will be accurate to some 50ft (15m) in position and to within 3-5 cm/sec in velocity.

Meteorology and Geodesy

The types of meteorological data required for civilian and military purposes differ quite substantially, so the United States operates separate satellite systems for each. Essentially, civilian requirements call for data appropriate for weather forecasting while military requirements are much more immediate.

The DoD's primary weather information system, the Defense Meteorological Satellite Program (DMSP), involves two satellites known as Block 5D-2s maintained in circular orbits at an altitude of about 530 miles (850km) and inclined at 98°. This orbit is sun-synchronous so that the angle between the orbital plane and a line from the Earth to the sun remains constant. The orbital period is about 101 minutes and the geometry of this orbit and the Earth's rotation works out such that each satellite views the whole surface of the Earth twice a day.

The sensor suite on the Block 5D-2s is very comprehensive and the data collected can be transmitted on demand directly to small mobile receivers on land or on naval vessels. Optical and infra-red images can be produced in near real time with resolutions of 2,000 ft (600m), and a scanning infra-red radiometer meas-ures vertical temperatures in the atmosphere and water vapour content. A passive microwave temperature sounder sees through clouds to measure temperature from the ground up to altitudes of more than 17 miles (28km), and an ultra-violet sensor measures the density of the atmosphere at altitudes of 60-150 miles (95-240km). An electron spectrometer helps predict auroral activity, other sensors monitor ionospheric conditions, and a gamma ray detector is carried to provide data about nuclear explosions.

Military applications

The information gathered is of enormous value. Strategic and tactical air missions can be planned to take account of prevailing weather conditions, and Strategic Air Command constantly updates ICBM guidance systems with relevant meteorological data which can affect a reentry vehicle's trajectory. Knowledge of conditions in the upper and lower atmosphere is also essential for the efficient planning of satellite reconnaissance missions, and ionospheric data allows the quality of high-frequency radio transmissions to be predicted.

Soviet meteorological satellites —

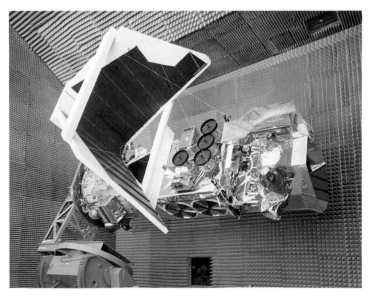

Meteor 2s — are in slightly lower orbits than their American counterparts, and two or three are kept operational. Resolution is believed to be slightly poorer than that of the Block 5D-2 but the sensor suite is broadly comparable, with optical and infra-red sensors and instruments for measuring the moisture content and vertical temperature profile of the atmosphere. Like American meteorological satellites, Soviet models can be interrogated directly by terminals anywhere in the world as the satellite passes overhead.

Geodetic satellites are used to produce maps of the Earth using photographic and radar techniques. They also provide data about the Earth's gravitational and magnetic fields. Among other things, this information enables the trajectories of ballistic missiles to be predicted accurately and is essential for the guidance systems of cruise missiles.

The US Defense Mapping Agency has launched a variety of satellites since the mid-1960s in its Geodetic Satellite Program and the Soviet Union has a similar programme.

Left: The Defense Meteorological Satellite Program (DMSP) involves two satellites designated Block 5D-2s. These are maintained in a circular orbit at an altitude of some 530 miles (850km) and inclined at 98°. This orbit is sun-synchronous.

Above: Photograph of Italy, Sicily, the Adriatic and part of the Mediterranean taken by the DMSP satellite using its operating line-scan system. The value of the system in assessing weather conditions is readily apparent in this image.

Satellite Warfare

IN ANY terrestrial conflict, satellites will play a key role. It is often said that satellites are 'force multipliers', meaning that they allow forces to be used more effectively and efficiently. However, as systems such as Navstar and Milstar become operational in the late 1980s space systems will become essential rather than just useful.

This increasing dependence on satellites has inevitably caused military planners to become concerned about the survivability of space systems as well as the merits of destroying an adversary's space assets. Consequently, steps are being taken both to increase satellite survivability and to develop anti-satellite weaponry.

The first requirement for any anti-satellite (Asat) weapon is a means to locate and identify an adversary's space systems. This is less demanding than might be imagined, since both superpowers have the necessary detection and tracking equipment already in place or under construction to fulfil other mission needs. The monitoring of satellites is already essential to avoid collisions, to prevent reentering satellites from

causing false attack alarms, and simply to determine what the other side is doing. Also, many satellite-tracking facilities exist as a by-product of the infrastructure intended to detect and monitor a ballistic missile attack.

Space debris

Tracking objects in space is, however, not an easy task. About 15,000 man-made items have been placed in orbit since the launch of Sputnik 1, among them rocket casings, dead satellites and various sorts of debris. Most have fallen back to Earth, but the American tracking systems still monitor about 5,600 objects. In addition to these known space hazards, incidentally, there are more than 10,000 pieces of debris — mainly the products of over 70 explo-

Below: It is vital for the USA and the USSR that their strategic command, control and communications (C^3) arrangements, which tie the systems together, are survivable. The NORAD Command Center is deep inside Cheyenne Mountain for just that reason.

sions in space — which are too small for the sensor network to track.

The nerve centre of the US space monitoring network is Cheyenne Mountain, near Colorado Springs, home of the North American Aerospace Defense Command (NORAD), which originally coordinated strategic air and missile surveillance systems and came to have responsibility for the surveillance and monitoring of all man-made objects in space. These space-related tasks are now handled by the USAF Space Command, whose Space Defense Operations Center is also housed in the NORAD Cheyenne Mountain Complex (NCMC) because many of the data collection and collation facilities are common to both commands. The NCMC itself is also a secure headquarters, buried beneath 1,500ft (500m) of granite and able to operate without outside supplies or power for up to 30 days.

The network of sensors which provides data on satellites, known as Spacetrack, includes systems dedicated to satellite monitoring as well as sources such as missile early-warning radars.

The first information about missile and satellite launches is provided by the Defense Support Program early-warning satellites described earlier. These produce initial tracking data for all rocket launches from Soviet and Chinese territory, and from the Indian, Atlantic and Pacific Oceans. Tracking is then handed over to the Ballistic Missile Early Warning System (BMEWS) which consists of three large radar installations at Fylingdales, England, Clear, Alaska, and Thule, Greenland.

BMEWS development

BMEWS was developed in the late 1950s and early 1960s, achieving operational status in 1962. Originally it was the primary means of obtaining early warning of a ballistic missile attack, but this function has been largely taken over by the DSP satellites and BMEWS is now more oriented towards attack characterization and the monitoring of satellite launches.

Each BMEWS site operates two different types of radar: a pulse-Doppler system with a fixed antenna measuring 400ft by 165ft (120 x 50m),

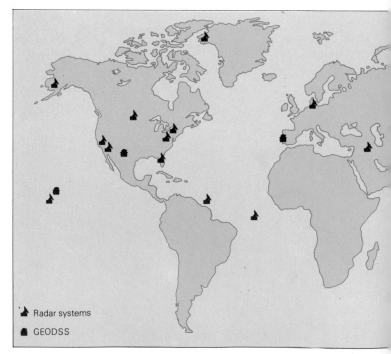

▲ Radar systems

■ GEODSS

Above: The crest of the United States Air Force Space Command, depicting Earth as seen from space. Activated on September 1, 1982, Space Command assumed responsibility for managing and operating assigned space assets, for centralizing space planning, and, with the advent of the SDI, for Air Force aspects of the strategic defence mission.

Left: The US Space Detection and Tracking System depends on a global network of radar and electro-optical stations, and includes stations set up for other purposes, such as the Ballistic Missile Early Warning System (BMEWS) sites at Thule (Greenland), Clear (Alaska) and Fylingdales (UK), as well as dedicated satellite tracking radars and the Ground-based Electro-Optical Deep Space Surveillance System (GEODSS) installations.

and a tracking radar employing an 82ft (25m) parabolic dish. In order to provide better target discrimination the computers at all three BMEWS sites were replaced in 1983 and modification of the radars was initiated. At Thule, new phased-array radars were scheduled to replace the older types some time after 1986.

Pave Paws radars

In addition to BMEWS, the US operates two Pave Paws phased-array radars, one at Otis Air Force Base, Massachussets, and another at Beale AFB, California, whose purpose is to detect and track SLBMs launched from the Atlantic and Pacific Oceans, and to provide satellite tracking data for Spacetrack. Pave Paws rather resembles a 100ft (30m) pyramid. The structure is in fact trapezoidal in plan with transmitter arrays on the sloping sides angled at 120° to each other. These combine to give 240° azimuth and 3° to 85° elevation coverage. The octagonal arrays measure almost 105ft (32m) across, and effective range is reported to be about 3,500 miles (5,500km). Construction of additional Pave Paws radars has started

at Goodfellow AFB, Texas, and at Robins AFB, Georgia.

Other radars providing data to both NORAD and Spacetrack include the Cobra Dane, FPS-85 and Perimeter Acquisition Radar (PAR) phased-array systems. Cobra Dane, located on a small rocky island at the western end of the Aleutians, less than 500 miles (800km) from the Soviet Union, became operational in 1977. Unlike Pave Paws, it has only one radar face, some 100ft (30m) diameter, which scans a volume of space spanning 120° in azimuth and about 80° in elevation. Its orientation

Below: A Ballistic Missile Early Warning System (BMEWS) site, with the radome acting as a visual as well as a radar beacon. Satellites over the Indian Ocean can sense the launch of Soviet ICBMs within 90 seconds of lift-off. They instantly warn ground stations at Guam and near Alice Springs in Australia; BMEWS then takes over to track and identify missiles directed towards North America and coming from the North. The BMEWS sites are thus an invaluable link in the strategic warning system.

is such that it covers Soviet missile test ranges, and it is sensitive enough to detect a grapefruit-sized metallic object at a distance in excess of 2,200 miles (3,500km). In its tracking mode it can simultaneously handle up to 200 objects at ranges of up to about 1,250 miles (2,000km).

The FPS-85, which became operational in 1969 at Eglin AFB, Florida, was originally designed as the main active sensor in the Spacetrack network, but is also performs missile detection tasks. The transmit and receive arrays are mounted side by side and are both inclined at about 45° to the horizontal.

PAR coverage

The PAR located in North Dakota was originally used in the US Safeguard missile defence system which was deactivated in 1976. Only the PAR remained operational, modified to serve as an additional element in the early-warning network. In this role it would validate data from other sources and help predict impact locations of incoming warheads. Its north-facing 120° coverage would also be particularly useful in the

Above right: A member of the 6th Missile Warning Squadron operates a Pave Paws missile warning console and monitors an oscilloscope display. The 6th MWS, located at Cape Cod Air Force Station, is tasked with providing warning and attack characterization of SLBM and ICBM attacks and surveillance, tracking and identification of space objects.

Right: The characteristic shape of the FPS-115 Pave Paws phased-array radar installation. Pave Paws operates at UHF frequencies and employs solid-state technology. The antenna array consists of two circular planar arrays, each some 100ft (30.5m) in diameter and consisisting of 5,400 discrete elements. The radar beams are electronically formed and steered, giving coverage of 85° in elevation and 240° in azimuth to a range of 3,000 miles (4,800km).

event of the loss of the BMEWS sites or in monitoring launches from the Arctic or Hudson Bay areas. The PAR's ability to track large numbers of missile warheads also enables it to monitor satellites and provide input to Spacetrack.

NAVSPASUR

The US Navy Space Surveillance (NAVSPASUR) system also feeds information into Spacetrack. NAVSPASUR is a multistatic continuous wave radar comprised of six receiving and three transmitting stations, all lying in a great circle extending from Fort Stewart, Georgia, to San Diego, California. The transmitter stations use large linear antennas to transmit a fan-shaped radar beam into space. The receiving stations detect radar reflections from satellites passing through the beams and can thus determine the orbital characteristics of any object traversing the beams between altitudes of 45 and 4,650 miles (75-7,500km). A modernization programme is in hand to double the altitude capability. The NAVSPASUR system is centred at Dahlgren, Virginia, and this site has been designated as the new Alternate NORAD Space Surveillance Center (ANSSC).

Spacetrack also receives data from NASA's tracking network and

from other sources such as sensors at the Kwajalein Missile Range in the Pacific and radars in such diverse places as Turkey, the Philippines and Thailand.

In addition to all these multi-purpose facilities, Spacetrack incorporates a number of dedicated surveillance systems, including four Baker-Nunn telescopes located at Edwards AFB, California, and in New Zealand, South Korea and Italy. These telescopes are mounted to follow the apparent motion of the stars as the Earth rotates, so that the photographs they take show satellites as moving objects against the fixed star background. This system

Above: This huge FPS-85 phased-array radar is operated by the 20th Missile Warning Squadron, located in Eglin Air Force base, a few miles south-east of Fort Walton Beach in Florida. The unit's primary mission is to give warning and attack character-ization of an SLBM attack aimed at the USA and southern Canada and the antenna faces across the Gulf of Mexico. The array is mounted at 45° with the square transmit antenna on the left and the octagonal receive antenna on the right; beams are electronically steered in elevation (105°) and azimuth (120°).

was developed in the 1950s and is very useful for plotting stable satellite orbits but analysis of data can take 90 minutes so manoeuvring objects cannot be tracked in real time.

The new Ground-based Electro-Optical Deep Space Surveillance (GEODSS) system, on the ·other hand will provide real-time data. Using larger optics coupled with low-light-level television cameras, GEODSS avoids the need to process film since images are converted directly into electronic signals for computer processing and visual display. The system can monitor an object the size of a football in geostationary orbit. The GEODSS network will consist of stations at White Sands, New Mexico, Taegu, South Korea, Maui, Hawaii, on Diego Garcia in the Indian Ocean and in Southern Portugal: the first, in New Mexico became operational in 1983 and the last, in Portugal, will be completed in 1988.

GEODSS improvements

Each GEODSS site has two 39in (100cm) telescopes, each with a 2.1° field of view, and one 15in (38cm) auxiliary telescope with a 6° field of view. Plans were already in hand by 1985 to improve the system by incorporating infra-red sensors, compensating image devices and more sensitive cameras, and the development of mobile, airborne and space systems operating on similar principles was also in progress.

It is not possible to catalogue Soviet satellite monitoring facilities in detail. Undoubtedly, like their US counterparts, Soviet early warning and missile tracking assets are also used. for space surveillance purposes. The Soviet ballistic missile early-warning system includes a launch-detection satellite network, over-the-horizon radar, and a series of large phased-array radars.

The early-warning satellite system described earlier gives the Soviet Union the most advanced information about missile and satellite launches. This network is supplemented by two over-the-horizon radars located at the eastern and western edges of the Soviet Union whose coverage intersects over the US

ICBM fields. This system gives less precise data than the satellite system but both working together can provide more reliable data than either working alone. The utility of the over-the-horizon radars for monitoring satellite launches is open to question. According to US diagrams, these radars give good coverage of the central United States but not of the coastal regions where the civilian and military satellite launch centres — Cape Canaveral, Wallops Island and Vandenberg AFB — are located.

Right: A 39in (100cm) telescope of the newly-operational Ground-based Electro-Optical Deep Space Surveillance system (GEODSS). The Low-light Level Television (LLTV) set at the rear gives a real-time capability.

Below: The two GEODSS buildings located at the White Sands Missile Range, some 30 miles (48km) from Socorro. Data obtained from GEODSS is sent direct to the NORAD Space Surveillance Center.

The Soviet Union does, however, possess a fleet of over 100 intelligence-gathering ships. These are primarily devoted to collecting information about Western naval operations but some are frequently observed off the coast in the vicinity of American satellite launch sites. In addition, five Soviet vessels, named after dead Cosmonauts, are used specifically to support Soviet space activities and to provide satellite tracking data. A new vessel, the *Marshal Nedelin,* launched in 1983, is believed to be used for space support purposes but may also have some input into the Soviet satellite monitoring network. Another ship of this type was under construction by 1985.

Radar coverage of space above and around the periphery of the Soviet Union is extremely comprehensive, and includes 11 large 'Hen House' ballistic missile early warning radars located at six points around the Soviet Union. The 'Hen House' network became operational in 1965 and a new system with a greatly enhanced capability was under construction by the mid-1980s, comprising six new Pechora class phased-array radars. Five of these duplicate or supplement the coverage of the 'Hen House' network, while the sixth is sited at Krasnoyarsk where it faces roughly northeast, providing coverage over approximately 2,500 miles (4,000km) of Soviet territory. These radars will considerably improve Soviet missile and satellite tracking capabilities, as the controversy surrounding the Krasnoyarsk radar clearly shows.

The 1972 Anti-Ballistic Missile (ABM) Treaty permits the construction of early-warning radars on the periphery of national territory facing

outward, and the Krasnoyarsk radar fulfils neither condition. The Soviet Union, however, maintains that the radar's purpose is satellite tracking, not early warning, although it is virtually identical to the Pechora radar: the simple fact of the matter is that this variety of phased-array radar is excellent for both purposes.

Other tracking radars

Other radars associated with the Moscow ABM system also have some satellite tracking capability. The most useful of these is presumably the Pushkino phased-array system. This radar is about 100ft (30m) high and has four faces each about 500ft (150m) wide. Its purpose is to control ABM engagements and in form and function it resembles the North Dakota Missile Site Radar. The older 'Dog House' and 'Cat House' radars south of Moscow also form part of the Moscow ABM system and may also be used for satellite tracking purposes.

In addition to the satellite surveillance capability provided by early-warning and ABM facilities, the Soviet Union operates some tracking stations in foreign territory and, of course, its own extensive space programme necessitates satellite tracking facilities. Radio telescopes, ground-based signals intelligence collection systems and optical sensors similar to the American Baker-Nunn devices are all used for monitoring space.

Below: Launched in 1983, the Soviet ship *Marshal Nedelin* is 702ft (214m) long and has a displacement of 25,000 tons. She carries numerous space- and missile-associated electronic systems and antennas.

Anti-satellite Weapons

The technical demands of an anti-satellite (Asat) system are less severe than those of an anti-ballistic missile system. Ballistic missile warheads, for instance, present themselves as targets for 30 minutes at most, and they are hardened to withstand the rigours of reentry into the Earth's atmosphere. Satellites, on the other hand, follow largely predictable orbits — sometimes for years — and they are relatively fragile objects. Even so, satellite interception is no easy task. A satellite in an orbit with an altitude of 100 miles (160km) travels at about 17,000 mph (27,000km/h), and a satellite in a Molniya orbit travels at about 23,000 mph (37,000km/h) at its closest approach to Earth of about 200 miles (320km).

Direct physical interception is clearly a difficult feat at these speeds, though the lethal radius of an interceptor can be increased in various ways, including the use of explosives. The most potent of these are, of course, nuclear. However, satellites well beyond a nuclear weapon's physical destructive radius can be disabled by the electromagnetic pulse generated by its detonation, so this technique carries the risk of disabling some friendly satellites.

Below: This montage shows the various types of space and ground-based anti-satellite weapons systems, together with the methods of destroying satellite sensors vital to an enemy. Low-orbit satellites are shown being knocked out by Asat missiles, while high-orbit navsats are knocked out by a combination of missiles and directed energy weapons. High and geostationary orbit satellites are targets for laser or particle-beam weapons. Soviet Asats can currently hit only low orbit satellites but new developments may, by the early 1990s, threaten high orbit satellites. No comparable American high orbit anti-satellite system is planned in that time frame.

 NATO
Soviet

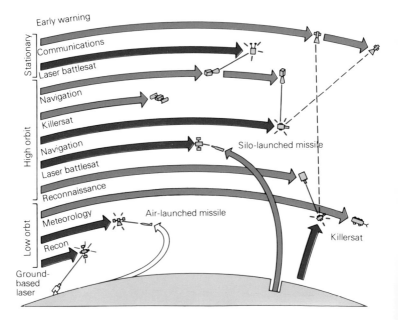

Another Asat technique is to transmit spurious command signals which could, for instance, cause a satellite to tumble irretrievably. Certain types of satellite also have further, specific vulnerabilities. Early-warning satellites, for example, could be blinded using lasers, and elint satellites could be jammed. In addition, an adversary can be deprived of his satellite assets by the destruction of the ground facilities which either control the satellite or receive data from it. These Asat techniques are available now, and in the future a variety of new technologies may be developed as a result of research into ballistic missile defence technology.

FOBS response

US interest in ASAT systems stemmed initially from a desire to stop the Soviet Union reaping the full benefits of its satellite assets in a conflict and was strengthened when the Soviet Union began to develop a Fractional Orbit Bombardment System (FOBS). FOBS was essentially a nuclear weapon placed into orbit by an SS-9 missile. Before it completed its first orbit, however, at about 500 miles (800km) from its target, retro-rockets were fired and it broke orbit to fall on its target. FOBS was seen as a particularly threatening system because it could attack from virtually any direction and could, therefore, circumvent north-facing US early-

Above: The Nike-Zeus missile was originally designed as an anti-ballistic missile (ABM) system. Armed with a 1MT nuclear warhead and guided to its target by ground radar, it had a very limited Asat capability, especially as its ceiling was only some 350 miles (550km).

warning systems. It was believed that it would be used in this way to destroy American ABM and early-warning radars, clearing the way for a full-scale ICBM attack.

The first American Asat programme was the satellite interceptor (SAINT), initiated in 1960 and intended to be placed into orbit by an Agena rocket stage on top of an Atlas-Centaur. SAINT would have inspected its target from an orbit alongside it and, if necessary, would have destroyed it with some unspecified form of interception device. This co-orbital technique was abandoned, untried, within three years in favour of another programme. The latter became operational in 1964 and was based on the Nike-Zeus missile originally designed as an anti-ballistic missile. Armed with a one-megaton nuclear warhead and guided to its target by ground-based radar, this system did not have to go through a complex series of orbital manoeuvres to intercept its target; instead, it was lofted directly into the path of its target satellite.

Above: The third test launch in the US Air Force anti-satellite (Asat) programme took place on September 13, 1985, high above the Western Test Range, off the Californian coast. The launch vehicle was a McDonnell

Douglas F-15 flown by the director of the Asat combined test force; the target was the Defense Advanced Projects Research Agency P78-1 gamma ray spectrometer satellite and the test was completely successful.

The Nike-Zeus system, however, had severe limitations, including a ceiling of only about 350 miles (550km), and was superseded by another system designated Program 437, which used the Thor intermediate-range ballistic missile to launch a nuclear warhead up to altitudes of about 800 miles (1,300km). It was judged necessary to detonate the warhead within about 5½ miles (9km) of the target in order to guarantee its destruction, and in tests (without a warhead) miss distances as low as one mile (1.6km) were recorded. In 1964, after three

out of four tests had proved successful, the system was declared operational; two missiles were maintained at Johnston Island in the Pacific Ocean, with supplies of extra systems held ready at Vandenberg AFB on the California coast, and despite some technical and financial problems Program 437 remained nominally operational until 1975.

The current US Asat programme, initiated in 1977 and scheduled to become operational around 1988, consists of a two-stage missile carrying a device known as a miniature homing vehicle (MHV) and designed

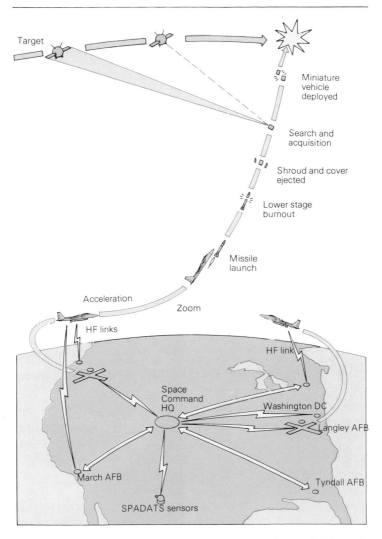

Target

Miniature vehicle deployed

Search and acquisition

Shroud and cover ejected

Lower stage burnout

Missile launch

Acceleration

Zoom

HF links

HF link

Space Command HQ

Washington DC

Langley AFB

March AFB

SPADATS sensors

Tyndall AFB

Above: The Asat mission operation concept, showing how the SPADOC inside Cheyenne Mountain will control and coordinate the use of the two squadrons of F-15s. After take off the F-15 will enter the launch area at high speed, accelerate and then zoom, prior to launching the Asat. Targets will be Soviet low-altitude reconnaissance and targetting satellites, which are considered the principal threat, rather than missile attack warning or navigation satellites.

to be launched from a slightly modi-fied F-15 aircraft. The first stage of the missile is the same rocket motor used on the Boeing Short-Range At-tack Missile (SRAM) carried by Strategic Air Command bombers, and the second stage is a Thiokol Altair 3 solid rocket, sometimes used as a fourth stage on the Scout rocket. The MHV sits on top of this assembly on a mounting which spins it up to 20 revolutions per second. The complete system is 17ft 9in (5.4m) long and 20in (50.8cm) in diameter and weighs about 2,600lb (1,180kg).

The MHV is a cylindrical device 12in (30.5cm) in diameter and 13in (33cm) long. The guidance system consists of eight infra-red telescopes; these lock on to the target and feed position data into an onboard computer which also keeps track of the MHV's speed and position by monitoring a laser gyroscope. Manoeuvring capability is provided by 56 single-shot rocket motors, mounted around the MHV's circumference and fired on command by the computer to nudge it onto an intercept course. The MHV carries no explosives: it is intended to ram into its target directly. With a collision speed of about 8 miles per second (13km/sec), the impact is roughly equivalent to a direct hit by a 16in shell.

Asat F-15 bases

Two squadrons of Asat F-15s are based at Langley AFB, Virginia and at McChord AFB, Washington, locations selected because of the orbital inclinations of Soviet satellites and the requirement that booster debris should fall over ocean areas. Using aircraft as the launch vehicle rather than ground-based missiles makes this Asat system enormously flexible: it is less vulnerable to attack, since the aircraft can be dispersed, and, being mobile, provides more launch opportunities than a fixed system.

Satellite interception missions will be coordinated by the Space Defense Operations Center (SPADOC), located in the Cheyenne Mountain Complex, which will select target satellites and calculate interception opportunities based on tracking data from Spacetrack and other facilities.

Right: An artist's concept of the USAF Asat Miniature Homing Vehicle (MHV) in the terminal phase of its attack on its satellite target, during the successful test on September 13, 1985. The MHV is 12in (30.5cm) in diameter and 13in (33cm) long and destroys the target by ramming (there is no explosive warhead), with the kinetic energy equal to a direct hit by a 6in shell. Steering is by means of 56 single-shot rockets.

The appropriate flight profile for an interception mission will be transmitted to the Asat aircraft's base and loaded into the aircraft's computer. The aircraft will then proceed to the missile launch point and just before launch will receive mission up-date information from one of the four Regional Operations Control Centers (ROCC) via HF radio data link. The flight profile will then be modified accordingly and the missile will be launched. The system is reportedly capable of intercepting satellites up to altitudes of about 620 miles (1,000km).

During the 1960s the Soviet Union demonstrated some of the techniques necessary to perform satellite interception, such as satellite manoeuvring and in-flight rendezvous, but the first full Soviet Asat test took place in October 1968. On October 19, 1968, Cosmos 248 was launched into an orbit with a perigee of 295 miles (475km) and an apogee of 337 miles (542km). The following day, Cosmos 249 was launched into an orbit with a perigee of 312 miles (502km) and an apogee of 1,018 miles (1,639km). Both satellites were in the same orbital plane. Within a few hours of the launch of Cosmos 249 close-range high-speed flyby occurred near the apogee of Cosmos 248's orbit, and immediately afterward Cosmos 249 exploded. On November 1, 1968, Cosmos 252 was launched and performed almost exactly as Cosmos 249 had done, again using Cosmos 248 as the target.

No further tests were observed for two years until an almost identical series of tests took place using Cosmos 373 as a target with Cosmos 374 and Cosmos 375 as interceptors.

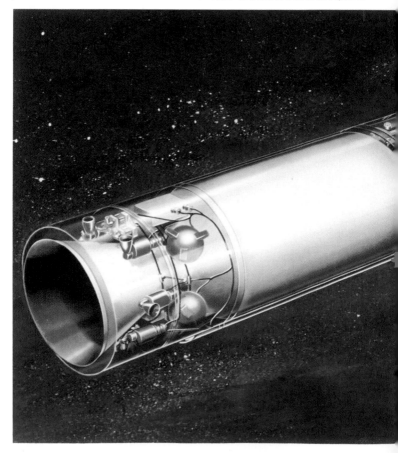

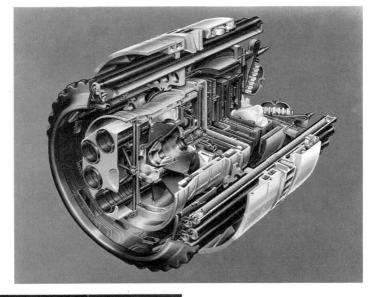

Above: The Miniature Homing Vehicle (MHV) of the US Air Force Asat. This cutaway shows the front end of the MHV with four of the eight infra-red telescopes in the yellow collar in the centre. These lock on to the target and feed positional data to an on-board computer situated immediately behind them. Around the circumference of the MHV are 56 single-shot rocket motors, seen here as long tubes. Extensive use is made of printed circuit boards (PCBs).

Left: A cutaway view of the complete Asat missile, which is 17.75ft (5.4m) long, 20in (50.8cm) in diameter, and weighs about 2,600lb (1,180kg). The front end (to the right of the picture) holds the MHV and behind it are two rocket motor stages. The second stage motor is a Thiokol Altair 3 solid rocket, as sometimes used in the Scout missile programme, while the first stage rocket motor at the rear of the missile is identical with that used in the Boeing Short-range Attack Missile (SRAM). The whole programme, using as many off-the-shelf components as possible, has been remarkably successful and trouble-free.

The three subsequent tests followed a broadly similar pattern except that the target satellites were launched from Plesetsk instead of Tyuratam, while the interceptors continued to be launched from Tyuratam. In addition, interceptions took place at altitudes ranging from 143 to 624 miles (230-1,005km).

Between the end of 1971 and early 1976 no full Asat tests took place, though there may have been some aborted tests. Nine tests occurred between February 1976 and May 1978, to be followed by a two-year break in testing while Soviet-American Asat negotiations were in progress. In 1980, after the suspension of negotiations, the tests recommenced and four took place before the Soviet Union declared a unilateral moratorium on ASAT tests in August 1983. No further tests had taken place by the end of 1985.

This Soviet system uses an SS-9 rocket to place an interceptor mechanism into orbit. This mech-

anism appears to be between 17ft 6in and 23ft 5in (4.5-6m) in length and about 5ft 10in (1.5m) in diameter with a weight of around 5,500lb (2,500kg), and can be equipped with an active radar or a passive infra-red/optical sensor, though the radar sensor seems to be more effective.

Two forms of interception have been observed. The more common involves placing the interceptor into an orbit which 'grazes' that of its target, and the interceptor launch is timed so that within one or two orbits the two objects are close to one another. The interceptor uses thrusters to approach the target and then explodes, sending a cloud of fragments, rather like a shotgun blast, toward the satellite. This technique can take several hours from launch to interception, ample time for a satellite to attempt to manoeuvre out of harm's way.

A slight variation of this method speeds the process up dramatically. The interceptor is placed into a low

Left: The Soviet Union's Asat capability predates that of the US by many years. Early Asat hardware is based on an interceptor launched by an SS-N-9 rocket. Initially, the mission lasted some hours, but a new direct approach method has reduced this to well under one hour. This artist's impression shows the Soviet interceptor approaching from the left, and in the process of exploding into a mass of destructive fragments.

parking orbit, beneath that of the target. The lower orbit means a shorter orbit time so the interceptor rapidly catches up to its target. At the appropriate point the interceptor fires its engines to pop up into the target's orbit and follow the normal homing-destruction sequence. Using this technique, the interval between launch and interception can be as short as 30 minutes, leaving little time for evasive action.

Asat operational

This co-orbital Asat has been considered operational since 1971, and up to 1981 the US Air Force judged tests successful if the interceptor passed within 3,280ft (1,000m) of its target. Since 1981, however, approaches within 5 miles (8km) have been judged successful. Whether this revision is due to the use of a better intercept mechanism or to a reappraisal of data is unclear. Overall, the system had demonstrated a 50 per cent success rate. Tests incorporating the radar sensor have had a 64 per cent success rate while all tests using the infra-red/optical sensor seem to have failed.

All attempted interceptions have taken place at altitudes of less than 1,250 miles (2,000km), though US estimates suggest that interception could be possible at altitudes of up to 3,100 miles (5,000km). As for the system's operability, the Tyuratam launch centre where the system is based is believed capable of launching 'several' interceptors per day.

In addition to this co-orbital Asat system, the 'Galosh' nuclear-armed ABM interceptors around Moscow may have the capability to attack low-altitude satellites.

As noted earlier, systems based on physical interception are not the only Asat possibilites, and one threat which satellites already face is electronic countermeasures. An adversary can attempt to interfere with a satellite's data and command links in an effort to throw a satellite off course or to jam signals transmitted to and from a satellite. Another approach is to illuminate a satellite with a ground-based laser, and some US satellites may already have been temporarily blinded in this way. Increased power might enable ground-based lasers to destroy satellites by causing them to overheat, or — eventually — by causing them physical damage. Two experimental Soviet ground-based lasers at Sary Shagan are already believed to have some Asat capability. Another possibility is to place lasers on high-flying aircraft to reduce atmospheric interference, though the power of such lasers would be limited.

Above: The scene at the Soviet Union's operational Asat base at Tyuratam. An Asat interceptor, atop a converted SS-N-9 rocket, sits ready on one of the launch pads, while further interceptors are brought forward on their launcher-erector wagons from the storage bay at the rear. Under operational conditions several Asat interceptors could be launched each day. The 'Galosh' nuclear-armed ABMs sited around Moscow may also have a limited ASAT capability.

Right: The scene at the Soviet Union's Research and Development establishment at Sary Shagan with a test on a ground-based laser taking place. Two such ground-based laser stations have been identified so far, and they are considered to have some Asat capability, even if only to cause the equivalent of jamming by interfering with satellites' sensors and data links. The application of greater laser power might lead to overheating and eventually to physical damage in the target. The USSR has poured vast resources into its military space programme.

The obvious environment for laser weapons is space, where the beam can propagate free from atmospheric attenuation and distortion. According to some reports, the Soviet Union is well on the way to building satellites equipped with lasers for use against other satellites, and feasibility studies have been conducted in the United States. Some assessments suggest that Soviet space-based Asat lasers could be deployed by about 1990.

A less exotic Asat weapon is the space mine. This concept calls for satellites armed with conventional explosives to be placed in orbit near their intended targets, and commanded to explode on the outbreak of hostilities.

Nuclear explosions in space have also been proposed as an Asat technique, and indeed, as we have seen, early American Asats, were nuclear armed; so is the Soviet 'Galosh' interceptor, which may have an Asat capability. ICBMs and SLBMs could be modified to place nuclear weapons in space in an Asat role, and nuclear weapons could also be used to arm space mines.

Nuclear-armed Asats would be effective in several ways. The fireball and intense radiation would destroy satellites in the immediate vicinity of an explosion, the electromagnetic pulse could upset the electronics on satellites thousands of miles away, and radio transmissions would also suffer disruption. Consequently, nuclear-armed Asat weapons are relatively indiscriminate and their effects would be felt by both hostile and friendly satellites.

Another means of degrading satellite systems is to attack their ground facilities. Options available include sabotage, conventional weapons, nuclear attack and high-altitude electromagnetic pulse.

All these potential threats are being taken so seriously that current satellite designs incorporate countermeasures against them. Solar panels and electronic systems are being developed which are more resistant to nuclear effects. Data and command links are being encrypted to avoid electronic spoofing while frequencies and transmission modes are being selected to provide resistance to jamming and EMP interference. Hardening against lasers is increasingly becoming a standard feature and satellites are also being given additional manoeuvring capability to evade interception. Also being studied is the use of stealth technology to render satellites less visible to radar. Despite these aids to survivability losses are clearly deemed inevitable, and spare satellites are being placed in orbits ready to plug any gaps which may appear.

Enhanced survivability

The survivability of ground stations is being improved by producing mobile facilities and efforts are also under way to increase satellite autonomy whereby a satellite can perform housekeeping functions such as correcting orbital anomalies independent of ground stations.

Further into the future, satellites may be fitted with more aggressive survivability aids such as lasers or homing projectiles intended to destroy approaching interceptors. The incorporation of so many survivability features in satellite design clearly indicates that Asat devices such as interceptors and laser weapons are deemed to be a genuine threat. This is not surprising, since both the United States and the Soviet Union have major research programmes in progress to explore new-technology space weapons, not just for Asat applications but also for defence against ballistic missiles.

Left: The advent of Asats means that satellites must now be designed for robustness rather than lightness. This picture shows the crater at Ground Zero following the Huron King underground nuclear test of a US satellite's vulnerablity.

Ballistic Missile Defence

THE NOTION of constructing defences against ballistic missiles is not new. In the early 1970s both the United States and the Soviet Union deployed anti-ballistic missile systems consisting of nuclear-armed missiles controlled from the ground under the direction of large radars. The American Safeguard system, using Sprint and Spartan missiles, was sited in North Dakota to protect ICBM fields, while the Soviet 'Galosh' system was centred on Moscow.

Many doubts were expressed about the effectiveness of these systems and there was particular concern about the operation of radars and control links in a nuclear environment. Consequently, the American system was dismantled in 1976, though the Soviet system continues to be maintained and upgraded.

Early progress

Research into ballistic missile defence (BMD) technology continued, and throughout the late 1970s and early 1980s there were reports that progress was being made. According to some experts, new technology could not only overcome the problems encountered by terminal defences such as Safeguard and 'Galosh' but might also provide the means for constructing much more comprehensive BMD systems.

In the West, matters came to a head in March 1983 when President Reagan made a major speech calling on US scientists to provide the means to make nuclear weapons obsolete. 'What', he asked, 'if free people could live secure in the knowledge that we could intercept and destroy strategic ballistic missiles before they reached our own soil or that of our allies?' Soon afterward the United States reorganized its BMD research programmes under the Strategic Defence Initiative (SDI). Research funds were doubled compared with the previously projected levels so that the SDI's funding for Fiscal Years 1985 to 1989 stands at $26 billion.

The Soviet Union, of course, does not provide details of its BMD programmes but a broadly similar effort

Above: The US Safeguard ABM system was developed in the late 1960s and the 1970s and involved two missiles, Sprint and Spartan. This was the first Sprint salvo being fired from Kwajalein Atoll in the Pacific against the nose cone of an ICBM launched some 4,200 miles (6,760km) away; it was a complete success.

Left: The US Army Safeguard Ballistic Missile Defence radar site near Grand Forks, North Dakota. This was declared operational in April 1975, but the whole system was closed down only a few months later. The Spartans were colocated with the radar; the four Sprint sites were 15 miles (25km) away.

to the SDI is in progress, possibly on a larger scale than its American counterpart.

The SDI is not geared toward constructing BMD weapons: its purpose — technology definition — is to investigate a range of BMD-related technologies and assess their potential. By the end of the decade, the knowledge gained will be available to identify the most promising technologies so that development of BMD weapons could begin if deemed technically feasible and politically desirable.

The research projects in the SDI are grouped into five major categories: attack monitoring, directed energy weapons, kinetic energy weapons, systems analysis and support programmes.

99

Attack Monitoring

The first requirement for defence against a ballistic missile attack is the ability to detect and track missiles with sufficient precision for them to be engaged by destructive devices, and to monitor the success or failure of each engagement so that, if necessary, interception attempts can be made. The technical description given to these tasks is surveillance, acquisition, tracking and kill assessment (SATKA), and investigations into SATKA technologies and concepts form one of the major elements of the SDI. Each phase of a ballistic missile's trajectory lends itself to particular SATKA techniques, though some technologies have applications in several or all phases.

Systems for use in the boost phase could have to monitor over a thousand Soviet missiles launched both from ICBM sites on land and from submarines in virtually any of the world's oceans. Heavy emphasis is placed on boost-phase SATKA because once it is over, within a few minutes of launch, the missile starts to dispense its warheads and decoys causing perhaps a ten-fold increase in the number of possible targets.

Launch detection

Launch detection is already a mature technology, and the demands of a surveillance system for boost-phase interception can be largely met by the improvements planned for early-warning satellites. Even so, additional systems will be required, because early-warning infra-red sensors actually lock on to a missile's ex-

Below: Any BMD system must be based upon a multi-layered space defence, with the layers tailored to the characteristics of each phase of a missile's flight. These phases are boost, post-boost, mid-course and terminal, and each lends itself to a particular form of Surveillance, Acquisition, Tracking and Kill Assessment (SATKA). The most profitable phases for the defence

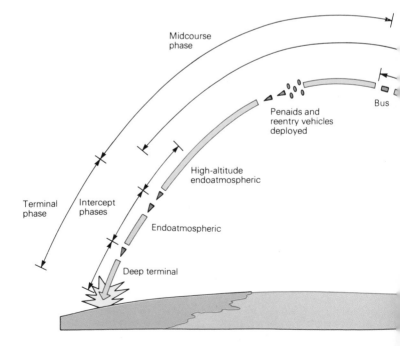

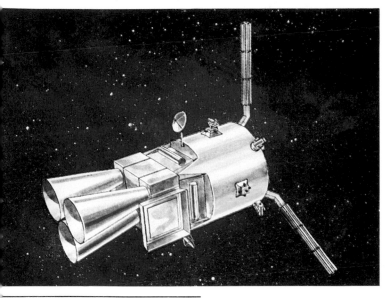

are the boost and post-boost phases when all the warheads are still concentrated in the nose of the missile, but this is also the most difficult because of the high speed of reaction required.

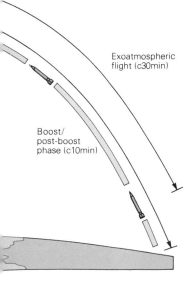

Exoatmospheric flight (c30min)

Boost/ post-boost phase (c10min)

Above: An important programme is the development of the Boost Surveillance and Tracking System (BSTS). Launch detection is already well developed, but the tracking of missiles during the boost phase is not so advanced. The BSTS programme is drawing on existing programmes which have been investigating the application of mosaic focal-plane arrays and advanced optics, as well as improving the data processing capability in order that the attack assessments can be passed direct to the user without the need for intermediate ground processors introducing delays.

haust plume rather than the missile itself. Infra-red sensors will probably be suitable for initial detection but for precise tracking and weapons pointing laser radars may prove to be more useful.

Even before the SDI programme was initiated, work was under way on improving boost-phase surveillance technologies by enhancing sensor performance and performing data processing onboard the satellite, so that attack assessment information could be provided directly to users without the need for ground processing stations.

A key programme intended to establish better boost-phase monitoring technologies, known as the Boost Surveillance and Tracking System (BSTS), draws on previously existing programmes investigating mosaic focal plane arrays, advanced optics and data processing. BSTS has both near-term and long-term goals: in the near term to enhance the performance of DSP early-warning satellites, and in the long term to establish the technology for boost-phase monitoring adequate for interception purposes. One of the programmes absorbed by BSTS is the High Altitude Low Observable (HALO) project which has already produced new infra-red sensors sensitive to longer wavelengths, new cryogenic devices and new lightweight optical devices.

Post-boost tracking

Tracking during the post-boost and mid-course phases poses a completely different set of problems. The number of objects to be monitored could exceed 100,000, and sensors would have to be able to distinguish warheads among a mixture of debris and penetration aids such as dummy warheads, chaff and aerosol masking clouds. Moreover, the warheads are obviously much smaller than the missiles themselves, and would be cold targets tracked against a deep-space background, as opposed to hot targets tracked against an Earth background.

Clearly, post-boost and mid-course SATKA will be enormously complicated but many promising ac-

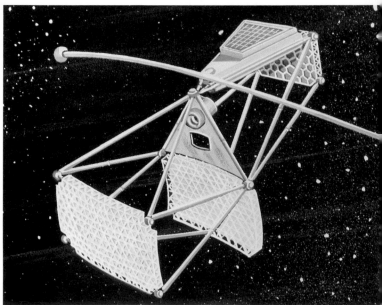

tive and passive sensors such as long wavelength infra-red and low-light-level devices are being studied. Phased-array microwave radars, ultra-violet laser radars, optical telescopes and various infra-red detectors are also being investigated and, depending on type, these sensors could be based in space, on aircraft, or on the ground. Other proposed detection techniques, which might be particularly good for target discrimination but which are more distant prospects, include lasers and particle beams powerful enough to induce variations in temperature or velocity which would reveal an object's nature.

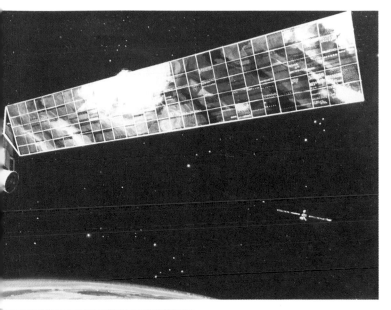

Above: One area of the research being conducted as part of the SDI programme is the development of high-altitude space-based sensors for boost surveillance and tracking. Another facet is a mid-course surveillance system that will be capable of detecting, tracking, and discriminating all objects in low Earth orbit, such as warheads, decoys and debris.

Left: Despite the impressions given in the Press, the SDI programme did not start from scratch, but has absorbed the many existing programmes, one of which was the High Altitude Low Observable (HALO) project. This had already produced new cryogenic devices, new infra-red (IR) sensors which are sensitive to longer wavelengths, and new lightweight optical devices.

It is intended to integrate many of the technologies being studied into a Space Surveillance and Tracking System (SSTS). This is envisaged as a sensor network which would monitor ballistic missiles in the mid-course phase as well as satellites. The SSTS, an independent SATKA asset, would accept data handed over by boost-phase sensors, support mid-course interception systems, and hand over data to terminal-phase sensors. The system would also have to be capable of withstanding Asat attacks and interference from jamming attempts and nuclear explosions. SSTS may emerge as a satellite network employing several types of sensors, but the final configuration will depend on the outcome of relevant research projects.

Airborne Optical Adjunct

Another major programme intended to develop and test SATKA technologies for the late mid-course phase is the Airborne Optical Adjunct (AOA), a modified Boeing 767 aircraft equipped with infra-red and optical sensors, plus associated data processing, navigation and communications equipment. Flying at altitudes of up to 65,000ft (20,000m), the AOA's sensors mounted on top of the aircraft would be able to peer into space from a vantage point well

Above: If it is decided to go ahead with SDI the crucial task of identifying and tracking warhead targets might be done, at least in part, by an airborne sensor system. This is an optical device, which has just been undergoing wind-tunnel tests to evaluate the effect it would have on the aircraft's performance.

Left: A model of the Boeing 767 which is being modified for the US Army under the designation Airborne Optical Adjunct (AOA). The aim of this project is to assess the technical feasibility of using airborne sensors to detect, track and discriminate ballistic missile reentry vehicles (RVs) and then handing over the data to ground-based radars.

105

above most atmospheric disturbances. Acquiring targets tracked by the SSTS as well as any which might have gone undetected, the AOA would also be able to hand over data to terminal SATKA systems. Estimates suggest that 14 patrol stations covered by a fleet of about 40 AOA-type aircraft would be able to provide full coverage of the United States, and as well as providing mid-course SATKA data the forward deployment of such systems could supplement boost-phase monitoring systems.

Attention is also being paid to mounting sensors aboard unmanned, long-endurance aircraft including such novel platforms as lighter-than-air vehicles and solar-powered RPVs.

Terminal tracking

During the reentry (or terminal) phase target discrimination problems are reduced because debris and light-weight decoys heat up or decelerate more rapidly than warheads. SATKA functions can again be performed by many types of sensor but the main candidates are airborne infra-red and optical sensors, and ground-based radar. Once a target was acquired by these detectors tracking data could be passed on to an interceptor's own sensors. Contracts were issued in mid-1985 for the development of phased-array terminal imaging radar (TIR) to detect and track incoming warheads, high in the atmosphere.

The design philosophy of SATKA systems intended for BMD use is to ensure that while each system can hand over data to the next, it can also function independently, so that neither loss of SATKA assets for a particular phase nor early detection errors would compromise the viability of other SATKA elements.

The development of SATKA systems relies on the integration of many specific technologies and projects which are not specific to any one phase. One of these is the Cobra Judy radar, based on a surface ship, which has recently been modified to improve its capability to monitor Soviet ballistic missile tests. The advantage of the ship-based radar is that it can get close enough to collect

data during the reentry phase of a test, providing information useful for the design and operation of both mid-course and terminal phase radars.

Other projects assess the emission characteristics of potential targets

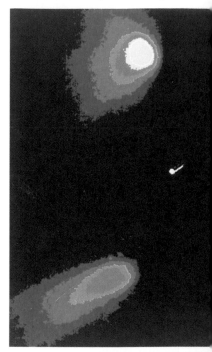

and their normal background to provide data bases for particular kinds of, for example, high-performance infra-red sensors, and several programmes are seeking to develop new cryocoolers, the refrigeration devices needed to maintain infra-red sensors at their extremely low operational temperatures. Still other projects are aimed at producing optical systems such as large, ultra-lightweight mirrors and the development of new sensing materials and fabrication methods.

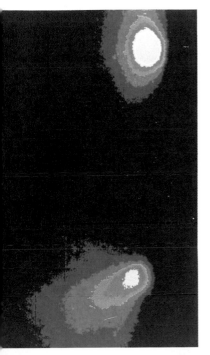

Left: One of the SDI studies is assessing emission characterisitcs (signatures) of potential targets. At the top are two ultraviolet signatures of a US missile in the boost phase, with its huge fire and smoke trail compressed by the atmosphere. Below are two infra-red signatures taken as the missile enters space, where, with no atmosphere to contain it, the trail spreads out.

Below: Part of the overall SDI project includes improvements to existing systems, such as this ship, USS _Observation Island_, which is fitted with the Cobra Judy radar system. The SPQ-11 radar in the huge rotating turret on the stern of the ship is used to collect data on Soviet strategic ballistic missile tests, and the advantage of a ship-based system like this is that it can get close enough to monitor the reentry phase of such tests.

Directed Energy Weapons

Directed Energy Weapons (DEWs) are seen as having considerable potential for intercepting ballistic missiles. DEWs would use high-power laser or particle beams to destroy their targets, but the relevant technologies are too immature for more than the general outline of the weapons to be determined. To cite just the most obvious technical challenges which must be overcome before DEWs become feasible, the power output of current directed energy devices is insufficient to destroy ballistic missiles over the necessary ranges, and beam focussing and aiming systems are inadequate for weapon purposes.

Enough is known about the candidate technologies, however, to categorize directed energy research programmes according to probable basing mode. Consequently, the DEW projects in the SDI are organized under four headings — space- and ground-based laser, space-based particle beam and nuclear-driven directed energy concepts.

Overlapping technologies

Not all the technologies being investigated under these headings are clearly distinct. Some are relevant to several DEW concepts and there is also some overlap between DEW and SATKA technologies. Nor does the current structure of research programmes exclude the prospect of variants on these themes: the emergence of, for instance, ground-based particle beam weapons as serious possibilities will depend on the outcome of various projects in progress.

One of the most mature systems for potential deployment in space is the chemical laser. Chemical lasers are powered by the reaction between two gases which also act as a lasing medium, and of several types under investigation the highest power levels have been achieved in those which use either hydrogen and fluorine or deuterium and fluorine.

A major chemical laser project known as ALPHA in progress at White Sands, New Mexico, is a hydrogen-fluorine laser which emits at a wavelength of 2.7 microns. The initial power of this system is expected to be two megawatts and it is designed to allow for the addition of more generator modules to increase power up 10 megawatts. Its purpose is to investigate the feasibility of building similar laser systems with outputs of 25 megawatts, large enough for use as a weapon.

Below: A highly developed technology, which is currently available and which might be suitable for deployment as part of a space-based SDI system, is the chemical laser. This is powered by the reaction between two gases in a combustion chamber, the chemical energy thus released being directly converted into light. The two pairs of reactants currently producing the highest power are hydrogen/fluorine, and deuterimum/fluorine. However, the light is in the infrared band and very large optical systems are needed to produce a focus over great distances. One current US project, known as ALPHA, is expected to produce 10 megawatts with add-on modules.

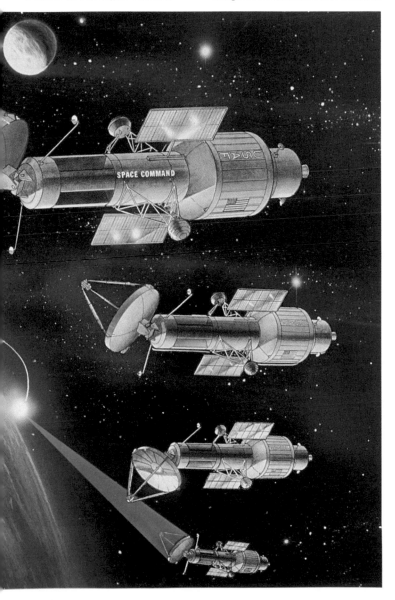

Also at White Sands is the Mid-Infra-Red Advanced Chemical Laser (MIRACL), a deuterium-fluorine laser. MIRACL is currently the most powerful laser in the Western world, with an output of 2.2 megawatts emitted at a wavelength of 3.8 microns. It is installed at the High Energy Laser System Test Facility (HELSTF) and is being used to develop a variety of DEW-related technologies including beam focussing and pointing. Eventually it is planned to test MIRACL's ability to shoot down sounding rockets.

LODE system

The space-based research programme also includes the Large Optics Demonstration Experiment (LODE) which will produce a 13ft (4m) optical system to demonstrate the feasibility of producing larger, high quality beam control devices. The LODE system will not be capable of handling powerful laser beams, however. Another major research area is in laser acquisition, pointing and tracking (APT). A central programme in this area known as Talon Gold was intended to produce a device capable of pointing a laser at a target extremely accurately. However, a pointing technique based on adaptive optics appears to have demonstrated a pointing capability beyond Talon Gold's goals, so this programme may be scrapped or re-oriented to support other elements of the SDI.

Right: The figure of a man puts this model of large cylindrical chemical laser into perspective. This is the ALPHA laser and the experimental test chamber, now being constructed at the San Juan Capistrano test site.

Below Right: The devastating effect of the Mid Infra-Red Chemical Laser (MIRACL) on the second stage of a Titan I booster missile body. In this test the target was under stress to simulate flight conditions.

Below: One possible configuration for a space-based chemical laser battle station for boost-phase interception. Such a station would have a 50ft (15.25m) diameter optical system and a power output of about 25 megawatts, with an all-up weight of about 100 tons (100,000kg). This may be a feasible proposition, but a much smaller station would be preferable.

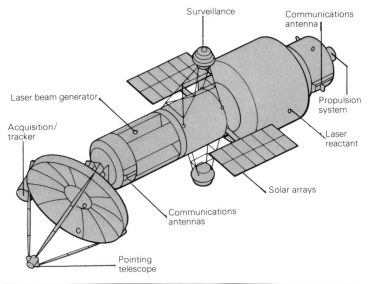

Surveillance

Communications antenna

Laser beam generator

Acquisition/ tracker

Propulsion system

Laser reactant

Solar arrays

Communications antennas

Pointing telescope

The space-based laser research effort contains many other projects intended to establish the concept's feasibility. Some are devoted to developing mirrors robust enough to cope with powerful laser weapons, yet light enough to be deployed in space. Such mirrors will probably be segmented since it appears to be easier to assemble several smaller mirrors than to construct a single large one. Others concern the problem of mirror contamination in space. Long exposure in space results in collisions with molecules and small particles which can cause significant damage to precision optical surfaces. Such damage could lead to the catastrophic failure of a mirror subjected to the powerful bolts of energy from a laser weapon. Still another line of research is aimed at improving the performance of chemical lasers by, for instance, enhancing the efficiency of the gas flow though the laser. Different forms of chemical laser, such as oxygen-iodine, and methods of damping spacecraft vibration are also being investigated.

While chemical lasers could probably provide the earliest DEW systems, excimer and free-electron lasers are judged to hold better long-term promise, though these will probably be much heavier and so are primarily envisaged as being ground-based.

Shorter wavelengths

The virtue of excimer and free-electron lasers is that they produce beams at shorter wavelengths than do chemical systems. Shorter wavelength beams do not disperse as much, so can employ smaller optical systems, and are absorbed better; their disadvantages are that they require higher quality optical surfaces, and the lasers themselves are relatively bulky and demand large power inputs so would probably have to be ground based.

The excimer laser derives its name from 'excited dimer', a dimer being a molecule consisting of two atoms. In this application, the dimer is made up of an inert gas (such as xenon or krypton) and a halogen (such as chlorine or fluorine). These can be

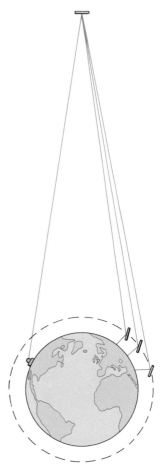

Below: Boost-phase interception using a ground-based laser firing a beam at a space-based mirror in geosynchronous orbit, which reflects it at an ICBM prior to warhead separation, is shown to scale here.

made to form a molecule by raising them to their excited states, for instance by passing an electric current through a mixture of the gases. When the dimer decays it emits a photon of light in the visible to ultraviolet range and 'cascades' of decaying dimers give rise to a coherent laser beam. The beam is emitted in pulses rather than as a continuous wave, and this may have some lethality advantages.

Several excimer laser projects are in progress with xenon-chlorine and krypton-fluorine devices receiving particular attention. Excimer lasers are less efficient than chemical lasers in terms of energy conversion, but a device known as a Raman cell allows arrays of excimer lasers to be coupled to produce one very powerful beam at a slightly longer wavelength than the input beam. A major project in this area is the Excimer Repetitively-pulsed Laser Device (EMRLD), which is intended to produce a megawatt class excimer laser before the end of the decade.

Below: The Mid-Infra-Red Advanced Chemical Laser (MIRACL) is currently the most powerful laser in the West, with an output of 2.2 Megawatts. It is being tested for a number of uses, one being anti-missile defence of ships at sea.

Bottom: The US Navy's laser beam director at the White Sands Missile Range. It contains an experimental pointing and tracking system which tracks targets in flight and then directs beams at selected aiming points.

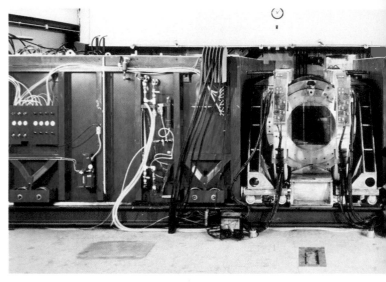

The free-electron laser works by passing a beam of electrons through a specially tailored magnetic field. This causes the electrons to oscillate which in turn causes them to emit photons of electromagnetic radiation. By adjusting the magnetic field, and by changing the energy of the electron beam, the emissions can be tuned to virtually any wavelength. Initially it was thought that free-electron lasers would only be suitable for ground basing, but some very powerful output levels have been achieved and there is some prospect that they may eventually be suitable for space basing.

The Lawrence Livermore Laboratory's Advanced Test Accelerator and the older Experimental Test Accelerator are being used for major free-electron laser experiments, and a variety of companies, government laboratories and universities are also conducting research into these devices.

Orbiting mirrors

The major thrust of research into excimer and free-electron lasers emphasizes their ground-based deployment with targets being engaged through a series of orbiting mirrors. The construction of suitable mirrors is a formidable task in itself because, as mentioned previously, their surface precision needs to be of a very

high order, and ground-basing raises the additional problem of atmospheric distortion. The latter problem is well on the way to being solved using a technique known as adaptive optics, which entails transmitting a low power laser beam from a beacon in space to a sensor near the weapon laser. The sensor measures the distortion in the beacon's beam and the weapon beam is pre-distorted accordingly so that as it passes through the atmosphere it reforms. Tests made at Maui Island in Hawaii bounced a ground-based laser beam off an aircraft 6 miles (10km) distant, and a system for compensating for atmospheric distortion performed very well.

Various adaptive optics projects are being undertaken, including research into sensors capable of monitoring wave distortion and methods of distorting laser mirrors accordingly. Associated with these efforts are studies into the effects of atmospheric phenomena on laser propagation.

Also receiving attention are the technologies associated with the construction of the space-based mirrors which would be needed for ground-based lasers to engage their targets. Some of the problems of constructing and controlling such mirrors are similar to those faced by space-based laser systems, but

Left: One solution to the problem of heavy lasers is to leave the laser on the ground and to use a space-based mirror to reflect the laser energy onto the target. Shown here is the high-power pulsed excimer laser experiment, known as EMRLD, which is investigating the ground-based element of such a system.

Below: Earth-generated pulsed laser beams being reflected by a space-based mirror toward high-altitude targets. This system would have many advantages, but one of the major difficulties is that the laser beam would be distorted and diffracted by the atmosphere. Methods have now been found, however, which should solve this problem.

others are specifically related to ground-based systems. The quality of reflective surfaces has to be higher, for instance, and the mirrors also have to be more robust, particularly for use with pulsed lasers. Research projects, therefore, are also devoted to resilient reflective coatings and highly efficient mirror cooling systems.

Particle beam weapons

Another concept receiving attention is that of space-based particle beam weapons. Particle beams of various kinds are being investigated but they all fall into two categories, neutral and charged particle beams. All particle beams start out as charged beams, accelerated up to speed and steered by magnetic or electric fields. To create a neutral beam, the charge is then removed by, for instance, using one of several techniques to strip one of the electrons from each particle.

In general, charged particle beams are unsuitable for long-range applications because they are deflected by the Earth's magnetic field and the beams tend to disperse relatively rapidly due to mutual repulsion of the particles. Particle beams also are ab-

sorbed by the atmosphere so ground basing is the favoured concept for DEW purposes.

There are two notable exceptions, however. It appears that a rapidly pulsed electron beam can be made to propagate though the atmosphere over respectable distances, as each pulse bores a hole through which the next can travel, and this technique may be useful for producing terminal defence systems. Another exception is that a negatively charged particle beam can propagate through a rarefied gas, following an ionized channel created by a relatively low power laser beam. The latter technique has been demonstrated and may permit charged particle DEWs to operate between altitudes of 50 and 375 miles (80-600km).

Below: This free-electron laser at the Los Alamos National Laboratory uses technology that was virtually non-existent until very recently. Such a system appears to be capable of producing visible light with efficiencies of as much as 40 per cent and is considered to be the optimum approach for a ground-based laser.

Below: A schematic diagram of a free-electron laser. In this device an electron beam is directed through a field-alternating ('corrugated') magnetic field. This forces the electron beam to follow a sinusoidal path (in effect, it is 'wiggled about') causing it to radiate energy and slow down. It is also possible to control magnetically the energy extraction rate and the colour of the light produced; ie, it can be 'tuned'.

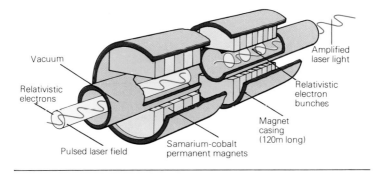

Vacuum
Relativistic electrons
Pulsed laser field
Samarium-cobalt permanent magnets
Amplified laser light
Relativistic electron bunches
Magnet casing (120m long)

Below: Return-wave system. Part of the beam is reflected, passed via a beam-splitter to a wave-form analyzer, and used to adjust the deformable mirror to pre-distort the outgoing beam. The atmosphere re-forms it.

Below: Outgoing-wave system. Each element of the deformable mirror vibrates at 10-40 Herz to modulate the outgoing beam. The reflected beam is analyzed and used to adjust the mirror to pre-distort the beam.

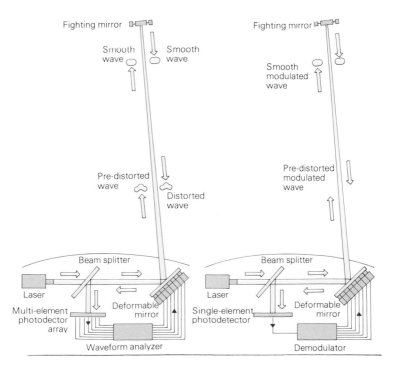

Fighting mirror
Smooth wave / Smooth wave
Pre-distorted wave / Distorted wave
Beam splitter
Laser
Multi-element photodetector array
Deformable mirror
Waveform analyzer

Fighting mirror
Smooth modulated wave
Pre-distorted modulated wave
Beam splitter
Laser
Single-element photodetector
Deformable mirror
Demodulator

Neutral particle beams appear to be more likely candidates for DEW applications. The basic technologies are being established by a variety of research programmes using facilities such as the White Horse Advanced Test Stand at the Los Alamos National Laboratory.

Particle beam weapons are a more distant prospect than laser weapons, but they do hold considerable long-term promise. For instance, they can disrupt a missile's electronic systems at energies much lower than those needed to cause physical destruction, and research is concentrating on methods of generating and steering sufficiently powerful beams, and on reducing the weight of the devices.

The X-ray laser

Another form of DEW system under investigation, the X-ray laser, is radically different from the others in the SDI. This weapon would operate by placing 50 or so lasing rods around a low-yield nuclear explosive. The rods would be pointed at hostile missiles and when the nuclear explosive was detonated the rods would emit intense pulses of X-rays toward their targets, destroying them either by a shock wave caused by the action of X-rays on the missiles' skin, or by the effect of X-rays on the missiles' electronics (the precise effect would depend on the frequency and intensity of the X-rays).

The exact nature of the X-ray laser is a closely-guarded secret, but at least two configurations have been suggested, each involving a nuclear weapon surrounded by lasing rods.

In one the lasing rods would be of dense metal and between 1m and 2.5m long; in another they would resemble carbon-fibre hairs 0.4in (1cm) long and one ten-thousandth that in diameter. Whether these descriptions are accurate or not, experiments conducted at the Nevada underground nuclear-test site have apparently been very encouraging.

How this type of weapon would be deployed is also uncertain. Permanent basing in space would violate the Outer Space Treaty which prohibits the placing of nuclear weapons in space, and some reports have proposed their installation on small missiles which could be launched in response to an attack.

Right: One method of destroying a ballistic missile would be to destroy its internal electronics, leading to a total malfunction of the missile itself, and one possible means of achieving this is with a neutral particle beam weapon. Shown here is the White Horse Advanced Test Stand, the second stage of a neutral particle beam device, which has produced a 100-milliamp beam using an RF quadropole accelerator.

Below: A schematic diagram of a possible neutral particle beam weapon for use in boost phase interception of ballistic missiles and mid-course interception of RVs. A full system might comprise 10 to 40 stations, each weighing 132,000lb (60,000kg).

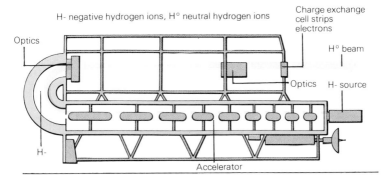

Kinetic Energy Weapons

Kinetic energy weapons are intended to destroy missiles by collision (kinetic energy is the energy of motion). Many different concepts are being evaluated, and it appears that such weapons could be employed to intercept ballistic missiles all the way from the boost phase until just before the warheads reach their targets.

One device receiving a great deal of attention is the electromagnetic rail gun. This works by passing a powerful electrical current — up to several megaamps — through two parallel rails. A projectile fits between the rails, making electrical contact across them, and as the current flows Lorentz forces act to propel it along the rails. There are many specific rail gun projects in progress and already some remarkable results have been achieved. One device has fired projectiles weighing 10½ oz (300gm) at speeds of 2.6 miles per second (4.2km/sec) and other achievements include firing low-mass projectiles at 25 miles per second (40km/sec) and firing a sequence of five projectiles in half a second. It is believed that it will soon be possible to increase projectile velocity to 60 miles per second (100km/sec) with a rate of fire up to 60 shots per second.

Basic operating modes

Two basic modes of operation are possible, one firing many unguided projectiles, the other firing fewer, heavier projectiles equipped with guidance systems, and both could be applied to interception from the boost through to the terminal phase.

Designing projectiles with guidance systems able to withstand the remarkable acceleration of launch

from an electromagnetic rail gun poses enormous technical problems, but a guided projectile has already conducted a pre-programmed manoeuvre after sustaining an acceleration in excess of 50,000g, and one promising approach may be to use a laser designator system based on the launch platform.

Two major rail gun projects are the ground-based hyper-velocity gun experiment (GBHE) and the space-based Sagittar. The GBHE will attempt to integrate several rail-gun technologies now being developed, while Sagittar will demonstrate the feasibility of using hyper-velocity guns in space and guiding their projectiles to intercept targets in space.

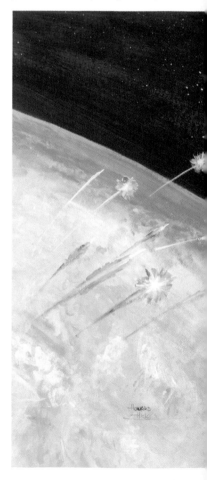

Right: An artist's impression of a space-based electromagnetic rail-gun firing hypervelocity projectiles, whose velocity may exceed 60 miles per second (100km/sec), at a rate of fire of up to 60 rounds per second. Experimental rail guns have already shown an ability to fire 3.5-ounce (2.5gm) projectiles at 5 miles per second (8.6km/sec).

Below: An electro-magnetic rail-gun concept where the bullet is a Lexan projectile with an aluminium skirt. A gas gun fires the bullet into the bore where the skirt makes contact across the rails and is vapourized by the current into a conducting plasma. Accelerating down the bore, it drives the bullet before it. Such systems could destroy ICBMs in all trajectory phases.

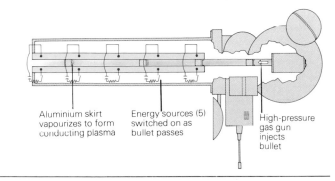

Aluminium skirt vapourizes to form conducting plasma

Energy sources (5) switched on as bullet passes

High-pressure gas gun injects bullet

Above: The US Army's Homing Overlay Experiment (HOE) flight-test vehicle streaks skyward from the Kwajalein Test Center to intercept a reentry vehicle from an ICBM launched from Vandenburg AFB, California.

Rail guns require enormous bursts of power for a fraction of a second to launch a projectile, and a key area of research is the study of systems which can satisfy these formidable requirements. The main device currently used is the homopolar generator, which consists of a heavy flywheel rotating extremely rapidly in a magnetic field. Completing a circuit across the homopolar generator converts the system's kinetic energy into electrical energy almost instan-

taneously. Other areas of research include the use of explosives to generate surges of electrical power, the production of robust materials for the rails and the design of projectile feed systems.

Kinetic kill vehicles

Another kinetic energy concept is the kinetic kill vehicle, essentially a small high-acceleration rocket equipped with a terminal guidance system similar in principle to the miniature homing vehicle used in the American Asat system. The concept involves a series of orbiting satellites, each fitted with many kinetic kill vehicles and with acquisition sensors carried either by these parent satellites or by remote battle managing satellites which would coordinate the sys-

Above: The HOE homing-and-kill vehicle closes at some 15,000ft/sec (4,600m/sec) on its quarry. The 15ft (4.6m) radial net has just unfurled.

Below: The HOE intercept mechanism. In the 1984 test accuracy was such that the axle hit the target, failing to test the efficacy of the weighted spokes.

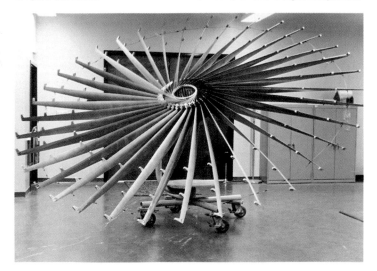

Above: In the 1984 test this video picture shows the rocket plume from the HOE homing-and-kill vehicle a fraction of a second before it hits the dummy warhead. The bar above the plume is a marker.

Above: One-tenth of a second after impact, this picture shows the debris as the steel net demolishes the RV in an absolutely spot-on hit. These pictures were taken through a 24in telescope at Kwajalein.

tem's operations. The number of satellites needed and the number of intercept mechanisms on each satellite would depend on the capability of the interceptors — range, weight, kill probability etc — and on the degree of protection required. One proposed scheme based on this concept, known as High Frontier, would comprise about 400 satellites each fitted with up to 50 interceptors. Several kinetic kill vehicle studies are in progress and the form of the projectiles is not yet established. One design goal is to construct a 44lb (20kg) projectile capable of accelerating up to 3.7 miles per second^2 (6km/sec^2), and research areas include appropriate guidance technologies and the effects of satellite sensor contamination by projectile exhausts.

Ground-launched interceptors

Ground-launched interceptors are also being investigated for midcourse and terminal phase applications. Such interceptors were developed in the late 1960s and early 1970s, but they were equipped with nuclear warheads to destroy incoming missiles, and future interceptors are expected to be accurate enough to use non-nuclear kill devices. Conventional explosives may be used, but their purpose would be to disperse fragments which would collide with incoming missiles.

Below: The US Army's Homing Overlay Experiment (HOE) as tested on June 10, 1984. An ICBM was launched from Vandenburg AFB in California and acquired during the boost phase by the

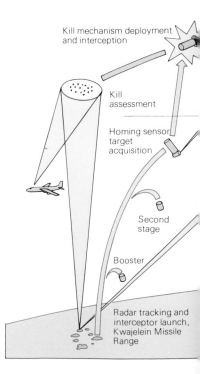

Kill mechanism deployment and interception

Kill assessment

Homing sensor target acquisition

Second stage

Booster

Radar tracking and interceptor launch, Kwajelein Missile Range

Above: Clouds of debris, following impact. The cloud at extreme left is all that remains of the target RV; the larger cloud in the centre is the remains of the HOE vehicle. Interception was at an altitude of over 90miles.

Above: A short time later all that remains of the HOE vehicle are these two clouds of debris. The debris from the RV target has dispersed over an area some 25 miles (40km) in radius. The test was an outstanding success.

Kaena Point radar. The HOE rocket was launched from the Kwajalein Range, the manoeuvrable interceptor having an on-board detector and a steel net destructor that opened up like an umbrella (see page 123). This net had 36 flexible spines, each with three small but dense weights near their ends, but their effectiveness was not tested as the central axle struck the target.

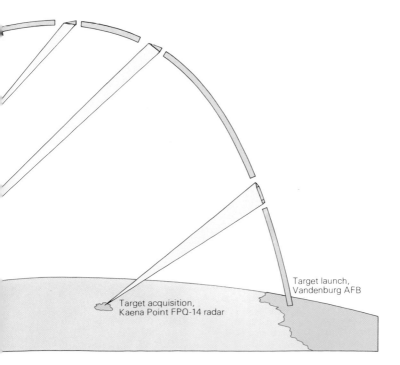

Target launch, Vandenburg AFB

Target acquisition, Kaena Point FPQ-14 radar

Ground-launched interceptors can be designed to destroy incoming warheads either inside or outside the atmosphere. The former are known as endoatmospheric and the latter as exoatmospheric systems, and research programmes into both forms are investigating the appropriate sensors, guidance systems, warheads, fuzing methods and propulsion technologies.

One major exoatmospheric interception programme, the Exoatmos-pheric Reentry Vehicle Interceptor Experiment (ERIS), will draw upon the many relevant research program-mes and will develop further the direct collision form of interception demonstrated by the Homing Overlay Experiment (HOE). Four HOE tests were conducted and the first three, while producing a great deal of valuable information, did not succeed in intercepting their targets, but in the fourth, conducted in June 1984, a reentry vehicle from a

Minuteman ballistic missile launched from Vanderberg AFB was destroyed by a HOE interceptor launched from Kwajalein Atoll. The interceptor unfolded weighted, umbrella-type arms about 8ft (2.5m) long to increase its destructive radius, just before scoring a direct hit at an altitude of about 90 miles (150km) and a closing speed of almost 5.5 miles per second (9km/sec).

Two of the main projects in the endoatmospheric interceptor research programme are the Small Radar Homing Intercept Technology (SRHIT) missile and the High Endo-atmospheric Defense Interceptor (HEDI). The SRHIT missile, almost 10ft (3m) long and about 10in (25cm) in diameter, is being used to test guidance and homing systems applicable to terminal phase interception. The missile manoeuvres by firing small rockets mounted around its circumference, in a manner similar to that of the Asat miniature homing vehicle. HEDI is envisaged as a higher altitude interceptor, capable of destroying reentry vehicles at altitudes between 9 and 37 miles (15-60km). Whereas test flights of SHRIT missiles have already taken place, HEDI is expected to begin testing around 1989.

Last-ditch defences

In addition to these endoatmospheric systems, research is also being conducted on last ditch defences close to high value targets. Concepts include radar-controlled rapid-fire machine guns and cannon-launched swarms of up to 1,000 small projectiles. At most, the effective range of these systems would be a few kilometers.

Left: Launch of the Small Radar Homing Intercept Technology (SRHIT) missile. Some 10ft (3m) long and 10in (25cm) in diameter, this missile is used to test guidance and homing systems for terminal phase interception. The object is to strike incoming warheads within the atmosphere.

Below: The High Endo-atmospheric Defense Interceptor (HEDI) is intended to be capable of destroying incoming RVs at heights of between 9 and 37 miles (15-60km). The interceptor uses an infra-red sensor and here is being set up prior to a hypersonic flow test.

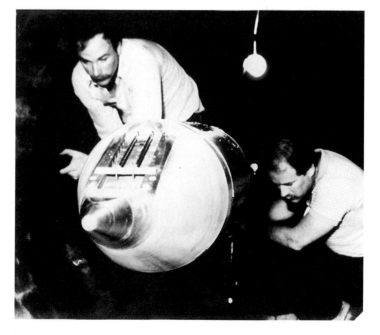

Systems Analysis, Battle Management and Support Programmes

One of the many research areas under these headings is the study of potential system architectures. Involving the definition of performance requirements, the derivation of estimates of the effectiveness of the elements of a defensive network, and analysis of trade-offs between various potential systems, this type of work requires the development of advanced modelling techniques and also investigates the form of potential Soviet countermeasures, the effects these would have on system performance, and methods of countering them.

Battle management is a crucial task which also requires extensive research. Any viable defensive network would need an extremely capable battle management system which would depend on remarkable data processing and command, control and communications (C³) techniques. The number of rapidly moving objects to be identified and tracked could exceed 100,000, and data concerning their probable function, position and velocity would have to be shuffled around the defensive network rapidly and reliably, even in the face of attempts to disrupt the system. This capability is well beyond the limits of current technology and would require dramatic improvements in, for instance, computer hardware and software, and communications technology. Conse-

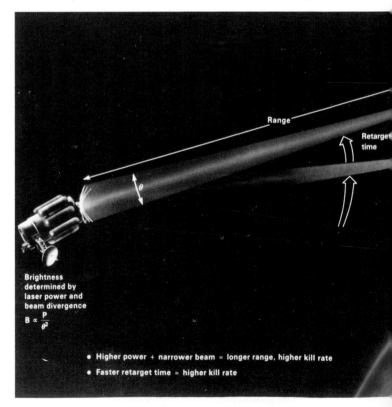

Range

Retarget time

θ

Brightness determined by laser power and beam divergence

$B \propto \dfrac{P}{\theta^2}$

- Higher power + narrower beam = longer range, higher kill rate
- Faster retarget time = higher kill rate

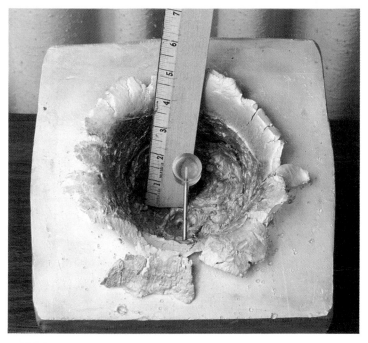

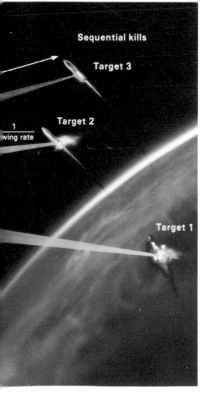

Above: The effect of firing a Lexan projectile (centre of picture) at a cast aluminium block from a rail-gun prototype.

Left: System kill rate depends on the brightness of the beam at the target and the time to swing from one target to the next.

quently, the SDI is funding research intended to produce much faster, more powerful and more reliable computer and communications technologies. ·

Research areas

Research includes, for example, projects in optical computers, laser communications, fault-tolerant computers, artificial intelligence, self-repairing systems and software-writing techniques. Despite the rapid rate of progress in these areas, many additional major advances will have to be made before battle management systems with the necessary capacity, speed, and resilience will be available. Indeed, some experts regard this area as one of the greatest technical challenges faced by the SDI.

Support programmes entail research into four main areas: survivability, lethality, power systems and logistics. Survivability research assesses the potential threat that defensive systems would face and promotes the development of tactics and technologies to minimize that threat. Specific projects in these areas include the study of hardening materials and techniques which would make systems more difficult to destroy. The results from this type of research are incorporated into system architecture studies.

Lethality research

Lethality research investigates the effects of destructive systems on a variety of targets, evaluating the sort of damage done by devices such as lasers, particle beams and projectiles to missiles, warheads and electronic systems. Such research is essential for assessing the design goals of candidate defensive systems and also relevant for survivability studies.

Space power systems research examines techniques for both generating and converting power to meet defensive system requirements. One of the main projects in this area is the development of the SP-100, a 6,600lb (3,000kg) nuclear reactor capable of generating 100kW, which should demonstrate the feasibility of constructing similar reactors with power levels up to about 2MW. Other projects include research into thermo-electric materials, heat rejection, explosives, rocket-driven generators, energy storage and nuclear fuels. The thrust of research is to design light, powerful energy supplies which can tick over for years but can rapidly produce extremely high power outputs if needed.

Space logistics research examines technologies which could launch, assemble, and maintain a defensive network. A key goal is to reduce launch costs because these would comprise a major portion of the expense of a space-based defensive system. Advanced reusable launchers, propulsion technology, robotics and in-orbit construction are among the research areas.

Above: Many SDI technology programmes will require space-based power sources, some with considerable power outputs. The SP-100 is the major technology demonstrator in this area and is due to fly by the 1990s. It is nuclear-powered, weighs some 6,600lb (3,000kg) and is designed to give a power output of 100 kiloWatts. It should prove the feasibility of similar reactors with outputs of up to 2MW.

Left: The first lethality test in the SDI programme was performed at the High Energy Laser System Test Facility (HELSTF) at White Sands on September 6, 1985. The test vehicle was the second stage of a Titan I missile booster (the cylindrical body in the centre of the picture) which was stressed mechanically to simulate one of the current Soviet missile systems. The target was irradiated by the MIRACL laser and was destroyed after a few seconds with the devastating results illustrated on page 111.

Soviet BMD Programmes

Although it is very difficult to obtain details of Soviet ballistic missile defence programmes, it is known that an effort broadly similar to the SDI is under way, and while in some areas the Soviet effort is greater and more advanced, in others the United States is ahead.

The Soviet Union has many programmes related to ground-launched interceptors, and since the late 1960s a network of 'Galosh' defensive missiles, designed to intercept ballistic missiles outside the atmosphere, has been maintained around Moscow. For many years only 64 missiles were deployed, although the 1972 ABM Treaty permits the deployment of up to 100, and in 1980 their number may have

decreased to 32. Since then, however, the number has steadily increased and at the end of 1985 had reportedly reached 100. In addition, the new Pushkino radar system, built to manage ABM engagements, is now operational, and other associated engagement and guidance radars are under construction.

New interceptors

There is some speculation about the nature of the interceptors in the network, though it is certain that a mixture of endoatmospheric and exoatmospheric missiles has been deployed. The former is the new quick-reaction SH-08, armed with a low-yield nuclear warhead, which would engage warheads very shortly

before impact, after combustion on entering the atmosphere had filtered out lightweight decoys and penetration aids.

In tests at Sary Shagan two SH-08 missiles were fired from the same silo within two hours, and it is believed that the ABM silos around Moscow can be reloaded, though at what rate is uncertain. The exoatmospheric 'Galosh' is also still incorporated in the Moscow defensive system, but reports differ about whether the network makes use of the new SH-04 interceptor. Similar to 'Galosh', the SH-04 is presumably more capable, and it may have the ability to stop and start its propulsion systems four or five times at very high altitudes, allowing the interceptor to loiter while ground radars sort out incoming warheads from decoys.

Left: The Moscow ABM system used the ABM-1B 'Galosh' exoatmospheric interceptor missile, shown here, for many years; this has a multi-megaton nuclear warhead and a range of some 186 miles (300km). Exoatmospheric interceptors are now silo-based and may be ABM-1Bs or the new SH-04.

Below: The second element in the new Moscow ABM system is the SH-08, a new atmospheric interceptor with a low-yield nuclear warhead. The SH-08 silos are reported to be reloadable; the rate is not known, but during tests at Shary Shagan two missiles were fired from the same silo within two hours.

The SH-08 and SH-04 are operated in conjunction with the 'Flat Twin' and 'Pawn Shop' radars to form a rapidly deployable defensive system designated ABM-X-3 by the US DoD. The 'Flat Twin' tracking radar takes its name from its two flat faces, while the 'Pawn Shop' guidance radar is so called because of its three spherical radar antennas mounted side by side. There is concern that the Soviet Union may covertly produce and store elements of ABM-X-3 in preparation for a sudden large-scale deployment of defensive systems.

SAM BMD applications

Several varieties of Soviet surface-to-air missiles may also be employed as ballistic missile interceptors. The SA-5 'Gammon', introduced in the 1960s and extensively improved since, can intercept targets up to altitudes of 18.5 miles (30km) and was repeatedly tested against ballistic missiles in the 1970s. The SA-10 is a newer, much faster surface-to-air missile able to reach a similar altitude and also thought to have some ABM capability. And still under development is the SA-X-12, which has also been tested against ballistic missiles. Although the tests in the ABM mode appear to have been against tactical ballistic missiles, the SA-10 and the SA-X-12 may have the potential to intercept some types of strategic ballistic missiles.

Soviet research into directed energy weapons is on a very large scale, with work proceeding on chemical, free-electron, excimer and

other laser systems, and on particle beams. At least six major DEW research facilities have been identified — three of them at Sary Shagan, an area long associated with Soviet ABM efforts.

Research into optical systems, beam control and power sources for DEWs is also proceeding, and although the scope of these activities is impossible to assess from the open literature, some developments indicate substantial progress. A rocket-powered generator has produced more than 15MW of power, and as yet has no counterpart in the West; power generating techniques using explosives are also thought to be well advanced, and some Western particle beam research is

based on the results of Soviet research conducted in the 1960s and 1970s. According to American estimates, some Soviet laser facilities already have an anti-satellite capability, and a prototype space-based anti-satellite laser may be produced before the end of the decade, with an operational system following during the 1990s. Components of a ground-based laser BMD system could be tested in the early 1990s, with actual deployment becoming a possibility around the late 1990s or after the year 2000, when space-based lasers will probably be another serious prospect.

Soviet research into kinetic energy weapons also parallels American research, though is probably hin-

dered by the Soviet lag in micro-electronics. Electromagnetic rail guns and more traditional interceptor techniques are being investigated.

As for other, related programmes, the Soviet Union's most serious deficiency is probably in the area of computer technology. The United States enjoys a substantial advantage in this area and, even so, is concerned about the advances necessary to support a comprehensive defensive network. The Soviet Union is, however, endeavouring to catch up in this field by research and by acquiring Western technology using a variety of legal and illegal methods. In the area of space logistics, the Soviet Union is better placed. Even though the American space shuttle is more advanced than Soviet launchers, the Soviet Union already has a very substantial launch capacity and is developing at least one type of reusable space shuttle as well as a new heavy-lift rocket.

Left: An SA-X-12 air defence battery in a tactical deployment. SA-X-12 has been tested against tactical ballistic missiles, and may also have some capability against strategic missiles.

Below: An SA-10 missile battery. With a ceiling of some 19 miles (30km) the fast reaction SA-10 may, like the SA-X-12, have a limited ABM capability.

Feasibility and Consequences

Although its intentions regarding future deployment of defensive systems are extremely difficult to judge, it is easy to understand fears that the Soviet Union may decide to enlarge upon its existing defences. The large phased-array radar under construction at Krasnoyarsk — described as an early-warning radar — is well placed to handle ABM engagements in defence of missile fields in the southern USSR, and the widespread deployment of surface-to-air missiles with at least some ABM capability is another cause of concern. Moreover, the development of the ABM-X-3 system suggests that a more effective, widespread BMD system may be fielded, possibly with very little warning, and even more capable defensive systems based on directed energy and kinetic energy weapons could eventually be deployed.

In the United States research into defensive systems was taking place even before the SDI came into being,

and some significant advances had been made. The Homing Overlay Experiment, for example, pre-dates the SDI, even though the first successful interception took place after the SDI had been announced. And the SDI itself will not produce any operational defensive weapons: the programme's goal is to investigate relevant technologies to support decisions in the early 1990s whether or not to proceed with the development of defensive systems. In terms of system feasibility, there can be no doubt that concepts for terminal defences offer the most immediate prospects of deployment, so that if the Soviet Union were to deploy large-scale terminal defences, the United States could follow suit.

Generally speaking, the earlier defences can intercept their targets, the more comprehensive will be the coverage they provide. In other words, terminal defences may be more readily constructed but the protection they offer is relatively limited.

Above: Multiple Independently-targetable Reentry Vehicles (MIRVs) being mated with the missile 'bus'. These MIRVs are for the Peacekeeper ICBM, but those fitted to other US and Soviet ICBMs and SLBMs are similar. One of the prime objectives in the SDI programme is to intercept strategic missiles during the boost phase, ie, before such RVs separate, as that way they kill many targets at once.

Left: The US government maintains that this large phased-array early warning and ballistic missile radar violates the 1972 ABM Treaty. Located at Krasnoyarsk, in Siberia, the radar is neither located on the periphery of the USSR, nor pointing outward, as required by the treaty. The USA says this radar closes the gap in the chain consisting of the 'Hen House' and large phased-array radars.

SPACE WARFARE

The technical obstacles which would have to be overcome by a comprehensive defensive network are clearly formidable, even if the nature of the threat remained unchanged. However, an adversary could take measures which would make BMD even more difficult. Missile boosters could be made to burn much more rapidly so that the boost phase — where the missiles are attractive, large and relatively soft targets — would occur almost entirely within the atmosphere, slightly reducing missile payload but severely curtailing the opportunities for boost-phase interception using particle beams, kinetic energy weapons and certain types of lasers. Boosters could also be made more reflective and could spin to disperse the effects of laser weapons, while protective coatings could be applied to a missile's surface and, of course, the number of missiles could be increased to place a heavier burden on defensive systems.

Post-boost complications
After the boost phase, a variety of sophisticated penetration aids such as reflective balloons and reflective gas clouds could be used, and warheads themselves could be disguised as decoys or designed to manoeuvre unpredictably, greatly complicating identification and tracking.

More aggressive countermeasures could be employed, such as blinding sensors with lasers and disrupting sensors and communications with nuclear explosives, along with a variety of electronic warfare techniques. The defensive systems of

Right: With the greatly increased accuracy of ICBM RVs, much attention is now being paid to the hardening of launch sites and to methods of disguising the actual location of land-based ICBMs. The US defence establishment went through a traumatic period while they tried to evolve an acceptable deployment mode for the MX missile system, one solution to which was the so-called 'racetrack' system. This picture shows a racetrack site hardness test.

one side would also be vulnerable to attack by those of the other and, particularly if beam weapons were used, such an attack could take place with little or no warning. And as well as constructing the means of overwhelming defences, an adversary would also have the option of circumventing them by means of bombers and cruise missiles.

In view of the complex nature of ballistic missile defence and the many opportunities for countermeasures, it is not surprising that there is considerable speculation about the feasibility of constructing effective large-scale systems. However, the notion of feasibility is often misunderstood. For instance, ballistic missile defence cannot be sim-

ply dismissed as unfeasible, because some systems will undoubtedly prove the contrary. The real question is whether such systems could provide worthwhile protection at a realistic cost. That, of course, begs another question: what constitutes worthwhile protection?

Extent of protection

No system is likely to provide complete protection: apart from anything else, it could never be realistically tested on a large scale. However, a defensive system capable of intercepting 90 per cent of its targets could have substantial merit. Clearly, the holocaust resulting from 1,000 incoming warheads would be catastrophic but a lesser one than that caused by 10,000, particularly in view of fears about the climatic changes that could be brought about by a large-scale nuclear conflict. Even a less effective defensive system could offer benefits to strategic stability. The prospect of intercepting even 20 per cent of an adversary's attacking warheads would introduce considerable uncertainty into an attack, since the attacker could not know which warheads would be destroyed and would therefore be unable to execute a well orchestrated first strike. In other words, it would be much more difficult to guarantee the destruction of an opponent's retaliatory forces, so the theoretical likelihood of either side attempting a first strike would be reduced.

Nevertheless, it would be foolish not to have reservations about the notion of deploying defensive sys-

Left: A time exposure photograph of eight Peacekeeper MIRVs passing through clouds toward an open ocean impact zone; in an actual atack the dispersion would be much greater. The use of multiple RVs on each missile has considerably enhanced the attacker's prospects of getting through to specific targets and has made the defence's task much more difficult. Nevertheless, a defensive system only 90 per cent effective would substantially reduce the ensuing devastation.

tems: while it may be possible to envisage deployments which would have beneficial effects on strategic stability, it is equally possible to envisage deployments which would have adverse consequences. For instance, an unrestricted race to deploy defences would lead to dangerous uncertainty and mistrust and could even lead to an attempt to strike before an opponent's systems were in place. Also, as we have already seen, defensive deployments could cause an arms race in offensive systems such as ballistic missiles, cruise missiles and bombers, resulting in, at best, even more expenditure on weapons with even more destructive capacity than that of today's arsenals and, at worst, in an utterly disastrous nuclear exchange. Furthermore, even if it is possible to imagine a stable strategic balance maintained by a preponderance of defensive systems, it is clear that the transition stage from today's balance of offensive systems would be fraught with difficulty.

Wider considerations

For the time being it is impossible to forecast the outcome of defensive deployments, but that outcome will certainly depend on more than merely technical considerations. The results of research will provide information about what forms of defence are available and at what price, but neither the feasibility nor the desirability of deployment will depend on only these factors. The planned nature of both the defensive network and offensive forces will have to be taken into account, as will the reactions of a potential adversary, and those reactions will depend on whether offensive and defensive deployments occur as part of a mutually agreed arms control arrangement or as part of an unrestricted arms race. And, ultimately, all these factors will depend on the strategic and political environment prevailing when the technical feasibility of deployment is established. In other words, the feasibility and desirability of deploying defensive systems will depend as much on decision-makers' political ingenuity as on scientists' technical ingenuity.

Manned Space Projects

If space-based ballistic missile defences are ever constructed, manned space missions will probably be indispensable for assembly and maintenance. Even without such defences, manned missions will be a major and probably an essential part of the growing military exploitation of space.

Until the advent of the space shuttle, the United States had not used manned space systems for military purposes to any great extent, though the 1960s a Manned Orbiting Laboratory (MOL) was mooted to perform reconnaissance activities and to in-vestigate the utility of manned space platforms for wartime command, control and communications. The system was to incorporate the Gemini space capsules mated with a cylindrical laboratory module. The programme began and a mock-up laboratory was launched in conjunction with an unmanned Gemini capsule modifed with a hatch in the rear to allow passage into the MOL, but the growing capabilities of unmanned reconnaissance satellites and the budgetary pressures of the Vietnam war led to the MOL's cancellation in 1969.

Below left: A shuttle vehicle at Space Launch Complex 6 at Vandenburg AFB, California. The first shuttle launch was scheduled for early 1986 but has been postponed, partly for local reasons but also because of the Challenger disaster at Cape Canaveral on January 28, 1986.

Below: A shuttle vehicle on the launch pad at Vandenburg. Both Vandenburg and Cape Canaveral are used for military and civil launches, but Vandenburg can be used to launch shuttles into polar orbit, which is not allowed from Cape Canaveral.

Some of the MOL's military experiments were transferred to the Skylab missions in the early 1970s and useful observations were made of activity at Soviet missile centres and along the Soviet/Chinese border. However, the United States remains unconvinced of the military utility of permanently manned military space stations for the time being, while the space shuttle can meet military needs for a manned space presence.

Military shuttle missions

The space shuttle was designed to meet both civilian and military needs, and approximately 30 per cent of its

SPACE WARFARE

activities are to be devoted to military purposes ranging from the launching of satellites to conducting space experiments. A fleet of four shuttles was constructed, and each vehicle was designed to be capable of about 100 missions. Mission duration is limited by fuel-cell life to about 10 days, but plans are in hand to extend this to about 18 days in the near future, and eventually time in orbit may be extended to 45 days. This could be achieved by docking the shuttle with an array of solar cells and batteries which would remain stationed in orbit.

The shuttle has a payload capacity of about 66,000lb (30,000kg) when placed into an orbit with a small inclination at an altitude of about 125 miles (200km). Launched into a polar orbit at the same altitude, its payload weight is reduced by about 50 per cent because it does not benefit so much from the additional velocity of the Earth's rotation. Naturally, higher orbits can be achieved with reduced payloads.

The shuttle provides unprecedented access to space, and as a launcher, experimental platform and in-orbit service system it will be used in-

to the next century. It has already demonstrated its military utility by launching a variety of military communications and elint satellites and by conducting laser tracking experiments, and in the future it will serve as a platform for many other SDI-related experiments, such as the testing of space-based sensors. The shuttle has also already demonstrated the ability to retrieve satellites for repair or refurbishment and to perform in-orbit maintenance, tasks beyond the scope of unmanned systems. In view of the enormous expense of military satellites this ability will be extremely valuable, and studies are already being conducted on the feasibility of modifying reconnaissance satellites for in-orbit refuelling and replenishment.

Shuttle Asat potential

Concern has been expressed by the Soviet Union about the shuttle's potential to pluck satellites from orbit. While this form of Asat activity is theoretically possible, it would be a dangerous enterprise, since booby-trapping satellites would be very simple. However, the shuttle could be used to inspect foreign satellites from a safe distance, and it could carry Asat weaponry, although cheaper unmanned systems would be more suitable.

A great deal of the scientific research using the shuttle also has relevance for the SDI. Flights of Spacelab — laboratory facilities designed specifically for the shuttle — have investigated how the shuttle disrupts space plasma; the wake created could aid in tracking large spacecraft and could affect the tracking of targets or the operation of beam weapons. The shuttle also glows slightly, a phenomenon of both scientific and military interest since it could reveal a spacecraft's position or interfere with sensitive instruments. Around 1990 the shuttle

Left: The space shuttle offers some inestimable advantages, one of which is the ability to recover satellites for repair or refurbishment. This has already been shown to be a cost-effective undertaking, and one that is beyond the capability of current unmanned systems.

Below: A maintenance man drops by to repair a faulty satellite, something that is currently possible only with the shuttle. The next step could well be in-orbit refuelling, replenishment and modification, and, ultimately, construction of space stations.

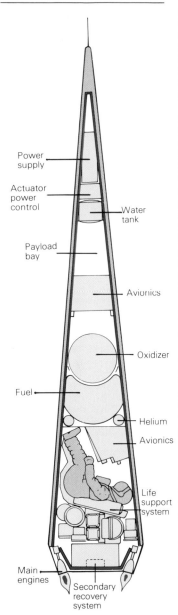

Power supply

Actuator power control

Water tank

Payload bay

Avionics

Oxidizer

Fuel

Helium

Avionics

Life support system

Main engines

Secondary recovery system

Above: A design for a single-seat Space Cruiser, with a synthetic outside view for the pilot. The ring of main engines would be fuelled by liquid oxygen in the spherical tank and liquid hydrogen from the larger tank immediately to the rear. Microthrusters would also be needed.

will be used to drop a probe on the end of a tether into a region of the upper atmosphere where more traditional sensor packages on sounding rockets can loiter only briefly. This will help design sensors and weapons for use against missiles passing through this region.

To support the military's launch requirements a shuttle launch pad has been constructed at Vandenberg AFB, California, a site which has been used for military and civilian launches of traditional rockets for many years. Vandenberg can be used to launch spacecraft into polar orbits, a practice not permitted from Cape Canaveral since it would involve the boost vehicle overflying the US mainland. The shuttle will be required to fly polar orbits, particularly for some military missions, hence the need for Vandenberg's Space Launch Complex 6.

Shared facilities

There is no absolute distinction between this military launch facility and NASA's civilian launch facility at Cape Canaveral in Florida; some civilian missions will take place from Vandenberg, and some military missions have already been launched from Cape Canaveral. The first shuttle launch from Vandenberg was scheduled to take place early in 1986, but even before the loss of *Challenger* there was a possibility of it being postponed until later in the year as a result of some problems with fuel handling facilities and conflicts with the civilian launch schedule.

Various supplements and successors to the shuttle are being put forward, though which ones will emerge as serious development possibilities remains a matter for conjecture. Small, single- or twin-seat space cruisers have been proposed for carrying out reconnaissance missions, placing satellites into orbit, repairing satellites, and for rotating the crews of permanent space platforms. Space cruisers might be launched by the shuttle, by MX-type rockets, or even from the back of a modified Boeing 747, and proposals suggest a capability for higher altitude operations than the shuttle can achieve so the vehicles would undoubtedly be useful, but whether they would be cost-effective is more open to question.

Various configurations of trans-atmospheric vehicle, a small manned spaceplane capable of taking off and landing using conventional runways, are also being studied. Its purpose would be quick-reaction, high-priority missions requiring only one or two orbits, possibly including reconnaissance or bombing, and the spaceplane could reach its target virtually anywhere in the world within 90 minutes. Studies are also in hand of a more advanced and fully re-usable successor to the shuttle. For launch, a jet-powered vehicle might take the space vehicle up to a high altitude and return to a conventional runway after separation. And the

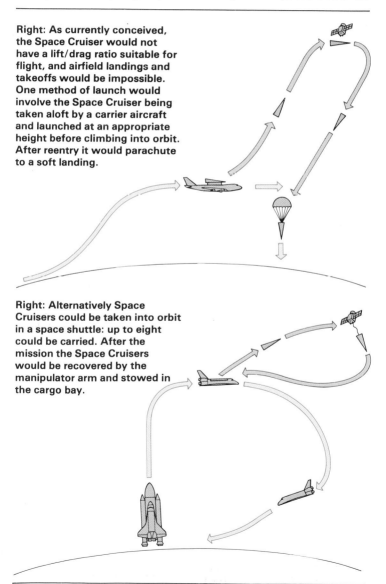

Right: As currently conceived, the Space Cruiser would not have a lift/drag ratio suitable for flight, and airfield landings and takeoffs would be impossible. One method of launch would involve the Space Cruiser being taken aloft by a carrier aircraft and launched at an appropriate height before climbing into orbit. After reentry it would parachute to a soft landing.

Right: Alternatively Space Cruisers could be taken into orbit in a space shuttle: up to eight could be carried. After the mission the Space Cruisers would be recovered by the manipulator arm and stowed in the cargo bay.

Above: USAF artwork shows a TAV riding on its expendable fuel (liquid oxygen/liquid hydrogen) tank carried aloft by an E-4B. The latter has a large dorsal fairing over the special communications systems for TAV missions.

Below: A Lockheed proposal for a transatmospheric vehicle (TAV) capable of operating at speeds ranging from subsonic to Mach 30. The TAV would be able to operate equally well on transcontinental routes.

United States may also build a new heavy-lift rocket capable of placing loads five times greater than the shuttle's into orbit.

A project much closer to realization is the US Space Station, a permanently-manned orbiting facility due to be in place in the early 1990s. The Space Station is an international venture and the DoD has no part in it, though in the future, particularly if large-scale BMD systems are deployed, the military may decide to acquire a similar station to provide such facilities as in-orbit repair and maintenance for other military space systems.

Soviet space stations

The Soviet Union already uses its Salyut space stations for both civilian and military purposes. The Salyut vehicles provide about 3,500 cubic feet (100m³) of working space, with docking ports at each end for manned capsules and unmanned resupply vessels and more than 20 portholes for visual observations and photography. Salyut 1 was launched in April 1971 and conducted a variety of

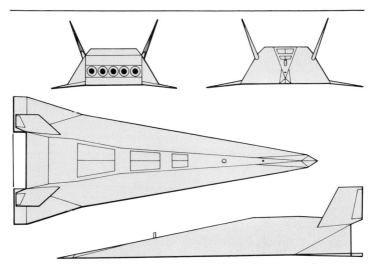

Above: A possible TAV design. The TAV programme involves many structural, material, propulsion and systems developments, but its shape is dictated almost entirely by aerodynamic and reentry considerations; as in all space vehicles the underside suffers the severest heating. Such a TAV would be capable of flight in and out of the atmosphere at up to Mach 25, and could travel to any point on Earth in two hours.

Right: One of many ideas for an
aerospace vehicle, this picture
shows a USAF shuttle-type
orbiter taking off on the back of
a reusable hevy-lift aeroplane.

Below: A Boeing concept of
what a permanently manned
space station might look like.
Everything shown is possible
using current technology.

scientific activities with the crew of Soyuz 11 on board. After a 24-day mission, Soyuz 11 returned to Earth but a faulty valve mechanism caused the capsule to decompress during reentry resulting in the deaths of the three-man crew. Salyut I remained unoccupied and finally reentered the atmosphere after 175 days in orbit. Salyut 2, launched in April 1973, broke up shortly after launch and

another attempt to launch a space station may have taken place about a month later, but the object, designated Cosmos 557, achieved only a very low orbit and burned up after 11 days.

In June 1974 Salyut 3 was launched, followed a few days later by Soyuz 14 which subsequently docked with the station. Salyut 3 was clearly a space mission intended

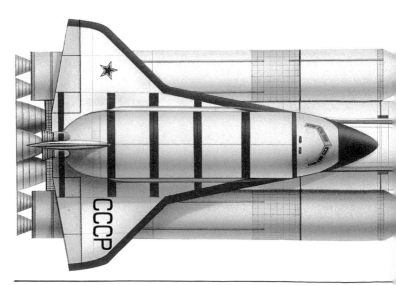

primarily for military purposes: its orbit was lower than that of its predecessors and better suited to reconnaissance purposes, and special targets were laid out near Tyuratam for the crew to photograph and establish the station's reconnaissance capabilities. After 16 days aboard the crew returned to Earth, leaving Salyut 3 to function automatically, and a docking failure with Soyuz 15 meant that the station remained unoccupied for the rest of its orbital life, though it continued to perform reconnaissance activities. It ejected a film capsule to Earth three days before burning up in the atmosphere after about seven months in orbit.

Salyut 4 was a civilian space station mission but Salyut 5, launched in June 1975, performed both military and non-military tasks. Again the primary military task was photographic reconnaissance, and the automatic film ejection system was used to return the film to Earth.

Left: A US conception of what a Soviet manned space station might look like. The space complex is based on existing Soviet hardware, while the shuttle vehicle coming in to dock is due to fly very soon.

Below According to reliable US sources this heavy lift launch vehicle, with a lift-off weight of some 1,968 tons, is now in the final stages of development. Although bearing some similarities to the US shuttle vehicle and its launcher, the Soviet system is not just a slavish copy: note, for example, that the main engines are on the large central body.

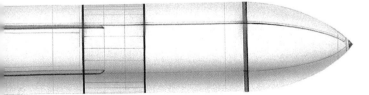

Salyut 7, launched in April 1982, appeared to perform both military and civilian functions, and as well as reconnaissance activities it seems that Salyut 7 has also been used as the target for laser tracking experiments. It is believed that the Soviet Union is developing a much larger space station with as many as six docking ports.

The Soviet Union is also developing a space shuttle which bears a remarkable resemblance to its US counterpart. The principal differences appear to be that the main engines are attached to the large central fuel tank, and the two strap-on boosters are liquid rather than solid fuelled. Two shuttle orbiters have been photographed by American reconnaissance satellites. Captive tests have been conducted with the shuttle carried on the back of a specially modified 'Bison' bomber,

but the system appears to have been delayed by booster problems.

A smaller, reusable 'mini-shuttle' may also be under development. Orbital tests have been conducted using a scale model, but it is unclear whether these are merely engineering tests or a step toward full-scale development. If it should be developed this mini-shuttle might be used for light cargo space station resupply missions and for quick-reaction military missions. The launch vehicle is expected to be the SL-X-16 rocket, also used as the strap-on booster on the larger Soviet space shuttle.

Finally, a large heavy-lift rocket currently under development is estimated to have a payload capacity of around 330,000lb (150,000kg) — about five times the American shuttle's limit — and may be used to launch space stations and, eventually, laser Asat devices.

Conclusion

THE SCENE IS SET for a dramatic expansion of military space activity. The capability of military satellites is growing and terrestrial military forces are becoming ever more dependent upon them. The militarization of space is frequently denounced by critics, but blanket condemnation is unjustified because many military satellites are useful for strategic stability and arms control. Elint satellites can provide advanced warning of increased military activity, while reconnaissance satellites can establish force structure and numbers, and can also monitor arms-control agreements. All this information is useful for building confidence and for preventing overreactions based on prudently pessimistic appraisals of a potential adversary's actions and capabilities. Consequently, although 'spies in the sky' seem sinister, those spies help to keep the superpowers both honest and calm.

Asat controversy

The weaponization of space is a different matter. Controversy rages about the effects of Asat on stability. Most people would agree that strategic stability would be enhanced if Asat weapons were not deployed. Unfortunately, technical realities preclude a simple superpower agreement not to deploy such systems. The problem is that many Asat systems — lasers, interceptors, ECM and so on — would be virtually undetectable until they were actually used, so an Asat treaty could prove unverifiable.

One solution occasionally put forward to resolve this difficulty is a ban on the testing of high altitude anti-satellite weapons to prevent the development of systems capable of threatening the most important satellites, which operate beyond the

Left: An exceptional picture, taken by a Royal Australian Air Force P-3, of the Soviet scale model shuttle vehicle being recovered from the Indian Ocean.

reach of current Asat weapons. The flaw is that the techniques developed for low-altitude Asat purposes could be adapted to high-altitude use basically by means of larger launchers.

An additional obstacle facing any future Asat treaty is the considerable overlap between Asat and BMD technology, which means an Asat agreement could either severely curtail ballistic missile defence projects or permit the development of new Asat weaponry under the guise of BMD projects.

So, despite the fact that a negotiated agreement on Asat systems would be desirable — and both superpowers can see the advantage of preserving their satellites' survivability — it would not be surprising if Asat development continued. It could be argued that the best way to protect satellites from hostile action is by posing a counter-threat to an opponent's satellites.

The situation regarding ballistic missile defences is even more complicated. The emotional appeal of constructing such defences is clear, but controversy surrounds both their technical feasibility and their strategic desirability. Research programmes now in progress will eventually resolve the technical debate one way or the other but the strategic questions will remain, simply because there are no definitive answers. One firm conclusion can be made, however.

Both superpowers are vigorously exploring BMD technology, yet, because it is more public, the American programme has attracted most criticism. But critics should bear in mind that, despite understandable reservations about actual *deployment* of BMD systems, while Soviet BMD research continues, American research should also continue as a hedge against future Soviet deployments or as a bargaining chip. If the Soviet Union maintains its BMD research programmes, it would be foolhardy for the United States not to do the same.

OTHER SUPER-VALUE MILITARY GUIDES IN THIS SERIES......

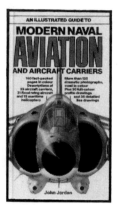

Air War over Vietnam
Allied Fighters of World War II
Battleships and Battlecruisers
Bombers of World War II
German, Italian and Japanese Fighters
  of World War II
Israeli Air Force
Military Helicopters
Modern Airborne Missiles
Modern Destroyers
Modern Fighters and Attack Aircraft
Modern Soviet Air Force
Modern Soviet Ground Forces

Modern Soviet Navy
Modern Sub Hunters
Modern Submarines
Modern Tanks
Modern US Air Force
Modern US Army
Modern US Navy
Modern Warships
NATO Fighters
Pistols and Revolvers
Rifles and Sub-machine Guns
World War II Tanks

＊Each has 160 fact-filled pages
＊Each is colourfully illustrated with hundreds of action photos and technical drawings
＊Each contains concisely presented data and accurate descriptions of major international weapons
＊Each represents tremendous value

If you would like further information on any of our titles please write to:
Publicity Dept. (Military Div.), Salamander Books Ltd.,
52 Bedford Row, London WC1R 4LR